A Quick Guide to API 570 Certified Pipework Inspector Syllabus

A Quick Guide to

API 570 Certified Pipework Inspector Syllabus

Example Questions and Worked Answers

Clifford Matthews

Series editor: Clifford Matthews

Matthews Engineering Training Limited
www.matthews-training.co.uk

WOODHEAD PUBLISHING LIMITED

Oxford   Cambridge   New Delhi

Published by Woodhead Publishing Limited, Abington Hall, Granta Park, Great Abington, Cambridge CB21 6AH, UK
www.woodheadpublishing.com
and
Matthews Engineering Training Limited
www.matthews-training.co.uk

Woodhead Publishing India Private Limited, G-2, Vardaan House, 7/28 Ansari Road, Daryaganj, New Delhi – 110002, India

Published in North America by the American Society of Mechanical Engineers (ASME), Three Park Avenue, New York, NY 10016-5990, USA
www.asme.org

First published 2009, Woodhead Publishing Limited and Matthews Engineering Training Limited
© 2009, C. Matthews
The author has asserted his moral rights.

British Library Cataloguing in Publication Data
A catalogue record for this book is available from the British Library.

Library of Congress Cataloging in Publication Data
A catalog record for this book is available from the Library of Congress.

Woodhead Publishing ISBN 978-1-84569-569-9   (book)
Woodhead Publishing ISBN 978-1-84569-684-9   (e-book)
ASME ISBN 978-0-7918-0289-2
ASME Order No. 802892
ASME Order No. 80289Q (e-book)

Typeset by Data Standards Ltd, Frome, Somerset, UK

Contents

Contents

SECTION III: NDE AND OTHER TESTING

Chapter 10: General NDE Requirements: API 570, API 577 and ASME B31.3

Chapter 11: The NDE Requirements of ASME V

The Quick Guide Series

The *Quick Guide* data books are intended as simplified, easily accessed references to a range of technical subjects. The initial books in the series were published by The Institution of Mechanical Engineers (Professional Engineering Publishing Ltd), written by the series editor Cliff Matthews. The series is now being extended to cover an increasing range of technical subjects by Matthews Engineering Publishing.

The concept of the Matthews *Quick Guides* is to provide condensed technical information on complex technical subjects in a pocket book format. Coverage includes the various regulations, codes and standards relevant to the subject. These can be difficult to understand in their full form, so the *Quick Guides* try to pick out the key points and explain them in straightforward terms. This of course means that each guide can only cover the main points of its subject – it is not always possible to explain everything in great depth. For this reason, the *Quick Guides* should only be taken as that – a quick guide – rather than a detailed treatise on the subject.

Where subject matter has statutory significance, e.g. statutory regulation and referenced technical codes and standards, then these guides do not claim to be a full interpretation of the statutory requirements. In reality, even regulations themselves do not really have this full status – many points can only be interpreted in a court of law. The objective of the *Quick Guides* is therefore to provide information that will add to the clarity of the picture rather than produce new subject matter or interpretations that will confuse you even further.

If you have any comments on this book, or you have any suggestions for other books you would like to see in the *Quick Guide* series, contact us through our website: www.matthews-training.co.uk

Cliff Matthews
Series Editor

How to Use This Book

This book is a 'Quick Guide' to the API 570 Certified Piping Inspector examination syllabus (formally called the 'body of knowledge' by API). It is intended to be of use to readers who:

- intend to study and sit for the formal API 570 Individual Certification Program (ICP) examination or
- have a general interest in the content of API 570 and its associated API/ASME codes, as they are applied to the in-service inspection of pipework.

The book covers all the codes listed in the API 570 syllabus (the so-called 'effectivity list') but only the content that is covered in the body of knowledge. Note that in some cases (e.g. B31.3) this represents only a small percentage of the full code content. In addition, the content of individual chapters of this book is chosen to reflect those topics that crop up frequently in the API 570 ICP examination. Surprisingly, some long-standing parts of the API 570 body of knowledge have appeared very infrequently, or not at all in recent examinations.

While this book is intended to be useful as a summary, remember that it cannot be a full replacement for a programme of study of the necessary codes. The book does not cover all the API 570 ICP syllabus, but you should find it useful as a pre-training course study guide or as pre-examination revision following a training course itself. It is very difficult, perhaps almost impossible, to learn enough to pass the exam using only individual reading of this book.

This quick guide is structured into chapters – each addressing separate parts of the API 570 ICP syllabus. A central idea of the chapters is that they contain self-test questions to help you understand the content of the codes. These are as important as the chapter text itself – it is a well-proven fact that you retain more information by actively searching (either mentally or physically) for an answer to a question than by the more passive activity of simply reading through passages or tables of text.

Most of the chapters can stand alone as summaries of individual codes, with the exception of the mock examination that contains cumulative content from all of the previous chapters. It therefore makes sense to leave these until last.

Code references dates

The API 570 ICP runs, normally, twice a year with examinations held in June and December. Each examination sitting is considered as a separate event with the examination content being linked to a pre-published code 'effectivity list' and body of knowledge. While the body of knowledge does not change much, the effectivity list is continually updated as new addenda or editions of each code come into play. Note that a code edition normally only enters the API 570 effectivity list twelve months after it has been issued. This allows time for any major errors to be found and corrected.

In writing this *Quick Guide* it has been necessary to set a reference date for the code editions used. We have used the effectivity list for the June 2008 examinations. Hence all the references used to specific code sections and clauses will refer to the code editions/revisions mentioned in that effectivity list. A summary of these is provided in the Appendix.

In many cases code clause numbering remains unchanged over many code revisions, so this book should be of some use for several years into the future. There are subtle differences in the way that API and ASME, as separate organizations, change the organization of their clause numbering systems to incorporate technical updates and changes as they occur – but they are hardly worth worrying about.

Important note: the role of API

API have not sponsored, participated or been involved in the compilation of this book in any way. API do not issue past

ICP examination papers or details of their question banks to any training provider, anywhere.

API codes are published documents, which anyone is allowed to interpret in any way they wish. Our interpretations in this book are built up from a record of running successful API 570/510/653 training programmes in which we have achieved a first-time pass rate of 75–80 %. It is worth noting that most training providers either do not know what their delegates' pass rate is or don't publish it if they do. API used to publish pass rate statistics – check their website www. api.org and see if they still do.

SECTION I: THE MAIN PRINCIPLES

Chapter 1

Interpreting ASME and API Codes

Passing the API ICP examination is, unfortunately, all about interpreting codes. As with any other written form of words, codes are open to interpretation. To complicate the issue, different forms of interpretation exist between code types; API and ASME are separate organizations so their codes are structured differently and are written in quite different styles.

1.1 Codes and the real world

Both API and ASME codes are meant to apply to the real world, but in significantly different ways. The difficulty comes when, in using these codes in the context of the API ICP examinations, it is necessary to distil both approaches down to a single style of ICP examination question (always of multiple choice, single answer format).

1.2 ASME construction codes

ASME construction codes (B31.3, B16.5, V and IX) represent the art of the possible, rather than the ultimate in fitness-for-service (FFS) criteria or technical perfection. They share the common feature that they are written entirely from a new construction viewpoint and hence are relevant up to the point of handover or putting into use of a piece of equipment. Strictly, they are not written with in-service inspection or repair in mind. This linking with the restricted activity of new construction means that these codes can be perceptive, sharp-edged and in most cases fairly definitive about the technical requirements that they set. It is difficult to agree that their content is not black and white, even if you do not agree with the technical requirements, acceptance criteria, etc., that they impose.

Do not make the mistake of taking the definitive requirements of construction codes as being the formal arbiter of fitness-for-service (FFS). It is technically possible, in fact commonplace, to use safely an item that is outside code requirements as long as its integrity is demonstrated by a recognized FFS assessment method.

1.3 API Inspection codes

API inspection codes (e.g. API 570) and their supporting recommended practice document (e.g. API RP 574) are very different. They are not construction codes and so do not share the prescriptive and 'black and white' approach of construction codes.

These are perhaps three reasons for this:

- They are based around accumulated expertise from a *wide variety* of equipment applications and situations.
- The technical areas that they address (corrosion, equipment lifetimes, etc.) can be diverse and uncertain.
- They deal with technical *opinion* as well as fact.

Taken together, these make for technical documents that are more of a technical way of looking at the world than a solution, unique or otherwise, to a technical problem. In such a situation you can expect *opinion* to predominate.

Like other trade associations and institutions, API (and ASME) both operate using a structure of technical committees. It is committees that decide the scope of codes, call for content, review submissions and review the pros and cons of what should be included in their content. It follows therefore that the content and flavour of the finalized code documents are the product of committees. The output of committees is no secret – they produce fairly well informed opinion based on an accumulation of experience, tempered, so as not to appear too opinionated or controversial, by having the technical edges taken off. Within these constraints there is no doubt that API codes do provide sound and fairly balanced technical opinion. Do not be surprised, however, if this opinion does not necessarily match your own.

1.3.1 Terminology

API and ASME documents use terminology that occasion-
ally differs from that used in European and other codes.
Non-destructive examination (NDE), for example, is normally referred to as non-destructive testing (NDT) in Europe
and the API work on the concept that an operative who
performs NDE is known as the *examiner* rather than the term
technician used in other countries. Most of the differences are
not particularly significant in a technical sense – they just
take a little getting used to.

In some cases, meanings can differ *between* ASME and
API codes (pressure and leak testing are two examples). API
codes benefit from their principle of having a separate section
(see API 570 section 3) containing definitions. These
definitions are selective rather than complete (try and find
an explanation of the difference between the terms *approve*
and *authorize*, for example).

Questions from the ICP examination papers are based
solely on the terminology and definitions understood by the
referenced codes. That is the end of the matter.

1.3.2 Calculations

Historically, both API and ASME codes were based on the
United States Customary System (USCS) family of units.
There are practical differences between this and the
European SI system of units.

SI is a consistent system of units, in which equations are
expressed using a combination of *base* units. For example:

$$\text{Stress}(S) = \frac{\text{pressure}(p) \times \text{diameter}(d)}{2 \times \text{thickness}(t)}$$

In SI units all the parameters would be stated in their base
units, i.e.

    Stress: N/m^2 (Pa)
    Pressure: N/m^2 (Pa)
    Diameter: m
    Thickness: m

Compare this with the USCS system in which parameters may be expressed in several different 'base' units, combined with a multiplying factor. For example, the equation for determining the minimum allowable corroded shell thickness of storage tanks is

$$t_{min} = \frac{2.6(H-1)DG}{SE}$$

where t_{min} is in inches, fill height (H) is in feet, tank diameter (D) is in feet, G is specific gravity, S is allowable stress in psi and E is joint efficiency.

Note how, instead of stating dimensions in a single base unit (e.g. inches) the dimensions are stated in the most convenient dimension for measurement, i.e. shell thickness in inches and tank diameter and fill height in feet. Remember that:

- This gives the same answer; the difference is simply in the method of expression.
- In many cases this can be easier to use than the more rigorous SI system – it avoids awkward exponential (10^6, 10^{-6}, etc.) factors that have to be written in and subsequently cancelled out.
- The written terms tend to be smaller and more convenient.

1.3.3 Trends in code units

Until fairly recently, ASME and API codes were written exclusively in USCS units. The trend is increasing to express all units in dual terms USCS(SI), i.e. the USCS term followed by the SI term in brackets. Note the results of this trend:

- Not all codes have been converted at once; there is an inevitable process of progressive change.
- ASME and API, being different organizations, will inevitably introduce their changes at different rates, as their codes are revised and updated to their own schedules.
- Unit conversions bring with them the problem of *rounding errors*. The USCS system, unlike the SI system, has never

adapted well to a consistent system of rounding (e.g. to one, two or three significant figures) so errors do creep in.

The results of all these is a small but significant effect on the form of examination questions used in the ICP examination and a few more opportunities for errors of expression, calculation and rounding to creep in. On balance, ICP examination questions seem to respond better to being treated using pure USCS units (for which they were intended). They do not respond particularly well to SI units, which can cause problems with conversion factors and rounding errors.

1.4 Code revisions

Both API and ASME review and amend their codes on a regular basis. There are various differences in their approach but the basic idea is that a code undergoes several addenda additions to the existing edition, before being reissued as a new edition. Timescales vary – some change regularly and others hardly at all.

Owing to the complexity of the interlinking and cross-referencing between the codes (particularly referencing *from* API *to* ASME codes) occasional mismatches may exist temporarily. Mismatches are usually minor and unlikely to cause any problems in interpreting the codes.

It is rare that code revisions are very dramatic; think of them more as a general process of updating and correction. On occasion, fundamental changes are made to material allowable stresses (specified mainly in ASME II-D, but also in ASME B31.3), normally as a result of experience with material test results, failures or advances in manufacturing processes.

1.5 Code illustrations

The philosophy on figures and illustrations differs significantly between ASME and API codes as follows:

- **ASME codes (e.g. ASME VIII)**, being construction-based, contain numerous engineering-drawing style figures and tables. Their content is designed to be precise, leading to clear engineering interpretation.

- **API codes** are not heavily illustrated, relying more on text. Both API 570 and its partner vessel inspection code, API 510, contain only a handful of illustrations between them.

- **API Recommended Practice (RP) documents** are better illustrated than their associated API codes but tend to be less formal and rigorous in their approach. This makes sense, as they are intended to be used as technical information documents rather than strict codes, as such. API RP 574 is a typical example, containing photographs, tables and drawings (sketch format) of a fairly general nature. In some cases this can actually make RP documents more practically *useful* than codes.

1.6 New construction versus repair activity

This is one of the more difficult areas to understand when dealing with ASME and API codes. The difficulty comes from the fact that, although ASME B31.3 was written exclusively from the viewpoint of new construction, it is referred to by API 570 in the context of in-service *repair* and, to a lesser extent, *re-rating*. The ground rules (set by API) to manage this potential contradiction are as follows:

- For new construction, ASME B31.3/16.5 are used – and API 570 plays no part.
- For repair, API 570 is the 'driving' code. In areas where it references 'the construction codes' (B31.3/16.5), these are followed *when they can be* (because API 570 has no content that contradicts them).
- For repair activities where API 570 and B31.3/16.5 contradict, then API 570 takes priority. Remember that these contradictions are to some extent false – they only exist because API 570 is dealing with on-site repairs, while ASME B31.3/16.5 were not written with that in mind. Two areas where this is an issue are:

- some types of repair weld specification (material, fillet size, electrode size, etc.);
- how and when pipework is pressure tested.

1.7 Conclusion: interpreting API and ASME codes

In summary, then, the API and ASME set of codes comprise a fairly comprehensive technical resource, with direct application to plant and equipment used in the petroleum industry. They are perhaps far from perfect but, in reality, are much more comprehensive and technically consistent than many others. Most national trade associations and institutions do not have any in-service inspection codes *at all*, so industry has to rely on a fragmented collection from oversea sources or nothing at all.

The API ICP scheme relies on these ASME and API codes for its selection of subject matter (the so-called 'body of knowledge'), multiple exam questions and their answers. One of the difficulties is shoe-horning the different approach and style of the ASME codes (B31.3/16.5, V and IX) into the same style of questions and answers that fall out of the relevant API documents (in the case of the API 570 ICP these are API 570/571/574/577/578). You can see the effect of this in the style of many of the examination questions and their 'correct' answers.

Difficulties apart, there is no question that the API ICP examinations are all about understanding and interpreting the relevant ASME and API codes. Remember, again, that while these codes are based on engineering experience, do not expect that this experience necessarily has to coincide with your own. Accumulated experience is incredibly wide and complex, and yours is only a small part of it.

Chapter 2

An Introduction to API 570

2.1 Introduction

This chapter is about learning to become familiar with the layout and contents of API 570. It forms a vital preliminary stage that will ultimately help you understand not only the content of API 570 but also its cross-references to the other relevant API and ASME codes.

API 570 is divided into nine sections (sections 1 to 9), four appendices (appendices A to D), three figures and four tables. Even when taken together, these are not sufficient to specify fully a methodology for the inspection, repair and rerating of pipework systems. To accomplish this, other information and guidance has to be drawn from the other codes included in your API document package. Figure 2.1 shows how the codes work together.

So that we can start to build up your familiarity with API 570, we are going to look at some of the definitions that form its basis. We can start to identify these by looking at the API 570 contents/index page. This is laid out broadly as shown in Fig. 2.2.

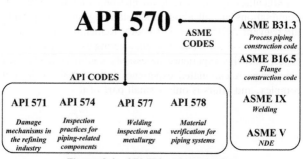

Figure 2.1 API 570: related codes

1 SCOPE

1.1 General Application
1.2 Specific Applications
1.3 Fitness-for-Service

2 REFERENCES

3 DEFINITIONS

4 OWNER/USER INSPECTION ORGANIZATION

4.1 General
4.2 API Authorized Piping Inspector Qualification and Certification
4.3 Responsibilities

5 INSPECTION AND TESTING PRACTICES

5.1 Risk-Based Inspection
5.2 Preparation
5.3 Inspection for Specific Types of Corrosion and Cracking
5.4 Types of Inspection and Surveillance
5.5 Thickness Measurement Locations
5.6 Thickness Measurement Methods
5.7 Pressure Testing of Piping Systems
5.8 Material Verification and Traceability
5.9 Inspection of Valves
5.10 Inspection of Welds In-Service
5.11 Inspection of Flanged Joints

6 FREQUENCY AND EXTENT OF INSPECTION

6.1 General
6.2 Piping Service Classes
6.3 Inspection Intervals
6.4 Extent of Visual External and CUI Inspections
6.5 Extent of Thickness Measurement Inspection
6.6 Extent of Small-Bore Auxiliary Piping and Threaded-Connection Inspections

7 INSPECTION DATA EVALUATION, ANALYSIS AND RECORDING

7.1 Corrosion Rate Determination
7.2 Maximum Allowable Working Pressure Determination
7.3 Retirement Thickness Determination
7.4 Assessment of Inspection Findings
7.5 Piping Stress Analysis
7.6 Reporting and Records for Piping System Inspection

Figure 2.2 The contents of API 570 (*continued over*)

8 REPAIRS, ALTERATIONS AND RE-RATING OF PIPING SYSTEMS

9 INSPECTION OF BURIED PIPING

APPENDICES

FIGURES

TABLES

Figure 2.2 (continued) The contents of API 570

2.2 Definitions

Some clarification points that you may find useful are as follows.

Section 3.1: alteration
Note how *alteration* is defined as a change that takes a piping system or component outside its usual design criteria envelope. What this really means is moving it outside the design parameters of its design code (ASME B31.3).

Section 3.4: authorized inspection agency
This can be a bit confusing. The five definitions (a to e) shown in API 570 relate to the situation in the USA, where the authorized inspection agency has some kind of legal jurisdiction, although the situation varies between states. Note this term *jurisdiction* used throughout the API codes (see how it is mentioned in definition 3.28) and remember that it was written with the states of the USA in mind.

The UK situation is different as the Pressure Systems Safety Regulations (PSSRs) form the statutory requirement. The nearest match to the 'authorized inspection agency' in the UK is therefore probably 'The Competent Person' (organization) as defined in the PSSRs. This can be an independent inspection body or the plant owner/user themselves.

When working towards the API 570 exam, assume that 'The Competent Person' (organization) is taking the role of the authorized inspection agency mentioned in API 570 section 3.4.

Section 3.5: authorized piping inspector
Again, this refers to the USA situation where, in many states, piping inspectors *have to be* certified to API 570. There is no such legal requirement in the UK. Assume, however, that the authorized piping inspector is someone who has passed the API 570 certification exam and can therefore perform competently the pipework inspection duties covered by API 570. It is also mentioned in definition 3.18.

Section 3.6: auxiliary piping

Be careful to differentiate the different types of piping recognized by API 570. These are important because the definition can affect the scope of inspection that is required (look briefly at API 570 section 6.6 and you will see the different requirements). These will be covered later; for the moment all you need to understand are the definitions.

Section 3.34: primary piping

This is the main piping system, > NPS 2 (i.e. greater than 2 inch nominal pipe size), which is not normally isolated and forms an essential part of the process circuit.

Other piping definitions

- **Secondary process piping** (see definition 3.40). These are ≤ (less than or equal to) NPS 2 pipes situated downstream of isolation valves, which are normally closed.
- **Small-bore piping** (see definition 3.41). These are ≤ (less than or equal to) NPS 2 pipes but may not be situated in locations that are normally isolated.
- **Auxiliary piping** (back to definition 3.6). These fulfil the definitions of *both* small bore and secondary process piping, i.e. they are ≤ NPS 2 and are normally isolated. Drains and vent lines are classic examples.

Section 3.10: defect

Note carefully what API 570 considers a *defect*. It is an imperfection that exceeds some code-acceptable criterion (or several criteria). Some components can contain surprisingly large imperfections (they may also be called *flaws*, *discontinuities* or *indications*) that are therefore, strictly, not classed as defects. It all depends on what the relevant code (e.g. ASME B31.3 for pipework or ASME VIII for vessels) says.

Remember: this definition of what is and is not a defect is far from being universal. It is, however, the one used by API 570 and its related US-based codes.

Section 3.11: design temperature
This rather complicated definition merely means that the design temperature of a component is the *maximum temperature* that it is designed to withstand at the same time that it sees the *maximum pressure* to which it is designed. This is normal design practice under virtually all common mechanical design codes.

Section 3.12: examiner
Don't confuse this as anything to do with the examiner who oversees the API certification exams. *Examiner* is the API terminology for the NDT technician who provides the NDT results for evaluation by the API-qualified piping inspector. API recognizes the NDT technician as a separate entity from the API authorized pipework inspector.

API codes (in fact most American-based codes) refer to NDT (the UK term) as *NDE* (non-destructive examination), so expect to see this used throughout the API 570 training programme and examination.

Section 3.21: MAWP
US pressure equipment codes mainly refer to MAWP (maximum allowable working pressure). It is, effectively, the maximum pressure that a component is designed for. European codes are more likely to call it design pressure. Simplistically, for the API 570 syllabus at least, you can consider the terms interchangeable.

Section 3.23: MT (and 3.29: PT)
US abbreviations are used throughout the API and ASME codes. The comparisons with those traditionally used in the UK are shown in Fig. 2.3.

Section 3.45: temporary repairs
You can get into a huge debate (with Health and Safety Authorities and others) about what constitutes a temporary repair, i.e. how long 'temporary' actually is. API 570 will not answer this for you. As with many codes, API simply says

Technique	UK common term	API/ASME term
The generic technique	NDT	NDE
Dye penetrant testing	DP	PT (penetrant testing)
Magnetic particle testing	MPI	MT (magnetic testing)
Fluorescent magnetic particle testing	MPI	WFMT (wet fluorescent magnetic testing)
Ultrasonic testing	UT	UT
Radiographic testing	RT	RT

Figure 2.3   NDE (NDT) abbrevations

that it is left to the authorized piping inspector to decide. For the purposes of passing the API 570 exam, don't think any more deeply than this: temporary means *not permanent* (according to the pipework inspector).

Sections 3.46: test points and 3.47: TMLs
TML stands for *thickness measurement location*, a chosen location at which thickness checks are to be carried out. The term TML is used throughout the API family of codes; note that it refers to a general location (e.g. an area of a pipework system such as a straight spool or a bend) rather than specifying *exactly* where the thickness test is to be done.

The exact area where a thickness test is to be done is known as a *test point* (definition 3.46). Even this is not absolutely precise, as the test point is in fact *an area* bounded by a circle of a certain maximum diameter (see Fig. 2.4).

Pipe diameter	Test point circle diameter
<10 inch (250 mm)	Maximum 2 inch (50 mm)
> 10 inch	Maximum 3 inch (75 mm)

Figure 2.4   Test point circle diameters

2.3 Owners/user inspection organization

Section 4.2: API authorized piping inspector qualifications and certification

Previously we saw the API view that the authorized piping inspector is someone who has passed the API 570 certification exam and can therefore perform competently the pipework inspection duties covered by API 570. Section 4.2 places the requirements for candidates to have minimum qualifications and experience, before they are allowed to sit the API 570 exams.

Section 4.3.1: responsibilities of user/owner

This section is quite wide-ranging in placing a raft of organizational requirements on the user/owner of a pipework system. This fits in well with the UK situation where the owner/user ends up being the predominant duty holder under the PSSRs. There is nothing particularly new about the list of requirements (listed as a to o); they are much the same as would be included in an ISO 9000 or UKAS accreditation audit. Note a couple of interesting ones, however.

Section 4.3.1(g): ensuring that all jurisdictional requirements for piping inspection, repairs, alterations and re-rating are continuously met

Remember that the term jurisdiction relates to the legal requirements in different USA states. In the UK this would mean statutory regulations such as the PSSRs, HASAWA, COMAH, PUWER and suchlike.

Section 4.3.1(k): controls necessary so that only qualified non-destructive examination (NDE) personnel and procedures are utilized

This means that API 570 requires NDE technicians to be qualified, although it seems to stop short of actually excluding non-US NDE qualifications. Look at section 3.12 and see what you think.

Section 4.3.1(l): controls necessary so that only materials conforming to the applicable section of the ASME code are utilized for repairs and alterations

This is clear. It effectively says that only ASME-compliant material can be used for repairs and alterations if you want to comply with API 570. This is currently difficult to comply with in the UK and other European countries where the Pressure Equipment Directive (PED) has a different system of material approval.

Therefore, accept that this part of API 570 does not fit well into the UK situation, but remember what API 570 actually says. The exam paper will be about what it says, not your view of how it fits into the UK inspection world.

Reminder: API 570 says that *only material conforming to the applicable section of the ASME code should be used for repairs and alterations.*

Chapter 3

More Advanced API 570

3.1 API 570 section 5

Section 5 of API 570 concentrates on inspection issues themselves. One of the key subsections is 5.3. Most definitions in this section are self-explanatory. Ones that are sometimes misunderstood are the following.

Section 5.3.1(a): injection point
Injection points are where a small-bore pipe (frequently containing high velocity fluid) enters a larger diameter pipe, normally at 90°, causing swirling and turbulence. Strictly, the following are not classed as injection points:

- pipe tee-pieces of similar or equal diameter;
- pipe Y-pieces of similar or equal diameter;
- orifice plates, control valves or reducers.

They may cause turbulence, which causes erosion, but they are not injection points according to API 570.

Section 5.3.1(g): environmental cracking
API 570 tries to explain this in section 5.3.7. It basically means stress corrosion cracking (SCC) in all its forms and a few other corrosion mechanisms such as hydrogen-induced cracking (HIC). The idea is that the corrosive mechanism is started off by the state of the environment. Note that this can be either the *internal* environment from the process fluid or the *external* environment, e.g. chlorides in the atmosphere encouraging SCC, particularly under lagging causing corrosion under insulation (CUI).

Section 5.3.1(h): corrosion under linings
API 570 explains this quite well in section 5.3.8. In this context, *linings* generally means coverings that are added after manufacture. These can be:

- stuck-on wrappings (rubber, epoxy tape, etc.);
- refractory linings (for fired components);
- loose cladding (e.g. stainless steel sleeves MIG or spot-welded to plain carbon steel pipe spools or valve bodies).

In general, bonded coatings such as paint, galvanizing, weld 'buttering', etc., are not considered in this category as linings.

Sections 5.3.1(j): creep cracking and 5.3.1(k): brittle fracture
These are well-known degradation mechanisms. Read the explanations in API 570 sections 5.3.10 and 5.3.11 respectively. They provide good explanations. You are unlikely to need to look at the referred standard API RP 920 mentioned in section 5.3.11.

Section 5.4: types of inspection and surveillance
The definitions here are core concepts of API 570. Note the five types listed below:

(a) Internal visual inspection
(b) **Thickness measurement inspection**
(c) **External visual inspection**
(d) Vibrating piping inspection
(e) Supplemental inspection

Read the definitions in sections 5.4.1 through to 5.4.5. Between them, they cover all types of inspections you are likely to encounter on piping systems. The most important ones are (b) and (c) shown in bold above. These are the ones referred to in section 6 of API 570 (have a look forward to API 570 table 6-1), which contains recommendations on maximum inspection intervals.

Simplistically, the whole concept of API 570 is to concentrate mainly on external visual inspections and thickness measurements, with other inspection types (a, d and e) being reserved for situations where the pipework inspector has a suspicion that something is wrong. This reinforces the overall API principle of reliance on an API-qualified authorized piping inspector.

Section 5.8: material verification and traceability
This section contains broad statements about the need for identification of materials used for pipework repairs or alterations. Note that there is a separate standard for this: API RP 578: *Material verification programme for new and existing alloy piping systems*. This is a required part of the API 570 syllabus coverage and is included in the API effectivity list. It will be covered later.

In concept, the API requirements for pipework repair/alteration material are simple:

- material must be capable of being identified and
- it must be the correct material for the job (i.e. ASME/API code compliant).

3.2 API 570 section 6.2: piping classes
The whole basis of API 570 is to divide piping systems into three classes. The classes are fairly specific to the petroleum industry and comprise a sort of a crude RBI (risk-based inspection) analysis. You will therefore see reference to classes 1, 2 and 3 pipework when looking at the recommended maximum inspection periods (API 570 table 6-1) and, as importantly, API 570 table 6-2 covering the recommended amount of CUI inspection.

3.3 Section 7: corrosion rate determination
API (and ASME) place great importance on the effects of wall thinning of pipes/vessels and the calculation of the maximum allowable working pressure (i.e. design pressure) that a corroded item will stand. To this end, they use various abbreviations and symbols to represent the various material thicknesses involved.

Note the following in API 570 section 7.1.1. The first two definitions are straightforward:

- t_{actual} is used to denote the actual thickness measured at the time of inspection.
- $t_{minimum}$ is a *calculated* value, rather than a measured one.

It is the minimum (safe) required thickness in order to retain safely the pressure and (more importantly) meet the requirements of the design code (e.g. ASME B31.3). This minimum thickness will normally include a corrosion allowance.

Then three general thickness readings are mentioned. These are less rigorously defined and are used to calculate corrosion rates with respect to time, i.e. long-term (LT) and short-term (ST) corrosion rates. They are all *actual* (measured) readings:

- $t_{initial}$ is the thickness measured at the initial inspection (not necessarily when it was new).
- t_{last} is the thickness measured at the last inspection (whenever that happened to have been). Think of this as the last inspection that was actually done.
- $t_{previous}$ is the thickness measured at an inspection previous to another specified inspection.

The timescale in Fig. 3.1 shows these in diagrammatic form.

Look at table 7-1 in your copy of API 570 (see Fig. 3.2) and note the following abbreviations and terms that have been used. The other terms and words used should be self-explanatory.

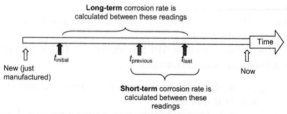

Figure 3.1   Thickness measurements

Term used in table 7-1	Its meaning
psig	Pounds per square inch (gauge)
NPS	Nominal pipe size (in inches)
Standard weight	This refers to 'standard weight pipe' – a US term
kPa	Pressure unit of kilopascal
S	Allowable stress for the material used in a pipe or pressure component (a value set by the material code)
E_l	A longitudinal weld efficiency factor
D	Pipe outside diameter

Figure 3.2   API 570 table 7-1 terminology

3.4   API 570 section 8: repairs and alterations

Section 8.1.3.1: temporary repairs
The terminology used here is straightforward. Note, however, the new acronym in this section. We will see it appearing regularly in the ASME codes:

- *SMYS* stands for *specified minimum yield stress.*

Note how it is a specified value (in the code). In practice, a material may be stronger than this specified minimum value.

Section 8.3 Re-rating
The word *re-rating* appears frequently throughout API codes. Re-rating of piping systems is perfectly allowable under the requirements of API 570, as long as code compliance is maintained. In the USA, the API authorized inspector is responsible for re-rating the pipework, once happy with the results of thickness checks, change of process conditions, etc. In the UK/European way of working, this is unlikely to be carried out by a single person (although, in theory, the API 570 qualification should qualify a pipe inspector to do it).

Re-rating may be needed owing to any combination of the following:

- corroded pipe (requiring reduced MAWP and/or temperature);
- an increase in process pressure/temperature conditions.

3.5 Section 9: Inspection of buried piping

Some terms used in this section that merit explanation are:

Above-grade (in section 9.1.1). This means above the ground level, i.e. ground level is known as 'Grade'.

Holiday survey (in section 9.1.3). The test carried out to find pinholes in a paint, epoxy, etc., or wrapped coating is known for some obscure reason as a *holiday test*. A high voltage (20 kV or more) is applied across the coating and any pinholes or cuts show up as a stream of sparks. Hence it is also sometimes called a *spark test*.

3.6 Appendix A: Inspection certification

API use quite a few API-specific terms in this section:

Body of knowledge (in section A.1). This means nothing more than the syllabus from which the API 570 certified piping inspector examination is compiled.

Certification (in section A.2). This refers to the API 570 certificate award that is made to candidates that sit and successfully pass the API 570 examination. It is awarded to individuals and has no connection with any other certificates that may be awarded to individuals or their employer company.

Recertification (in section A.3). This refers to the procedure whereby an inspector's API 570 qualification is revalidated every three years. Most recertifications are done simply by a CV assessment and the payment of a recertification fee to API. If, however, the inspector has not been involved in pipework inspection activities for at least 20 % of the time over the past three years, it may be necessary to resit the examination.

A recently introduced API rule requires API 570 inspectors

to sit an on-line update test every six years, to cover API 570 code changes that have been made since they passed the exam. These changes are usually small.

3.7 API 570 familiarization questions

Here are some questions to help familiarize yourself with the content of API 570. You can track down the answers from the code section given in the question. You will find the answers near the end of this book.

Q1. API 570 section 3.37: definitions
The replacement of a corroded pipe spool with one identical in all respects is known by API 570 as:

(a) A repair ☐
(b) A re-rating ☐
(c) An alteration ☐
(d) A code concession ☐

Q2. API 570 section 3.6: definitions
Which of these is defined as primary process piping by API 570?

(a) Piping that is isolated most of the time ☐
(b) Piping in normal active service that generally cannot be isolated ☐
(c) Vent and drain pipes ☐
(d) Piping downstream of normally closed block valves ☐

Q3. Section 3.41: small-bore piping: definitions
Small-bore piping is defined as piping that may not be situated in locations that are normally isolated and is equal to or smaller than:

(a) $\frac{1}{2}$ inch NPS ☐
(b) 1 inch NPS ☐
(c) 2 inch NPS ☐
(d) 3 inch NPS ☐

Q4. Section 3.12: The Examiner: definitions
Who is 'The Examiner' according to API 510:

(a) Any piping inspector ☐
(b) An API 570 certified piping inspector ☐
(c) Any NDE technician ☐
(d) An NDE technician with accepted NDE qualifications ☐

Q5. Section 3.21: MAWP: definitions

MAWP means much the same as?

(a) 90 per cent design pressure ☐
(b) 150 per cent design pressure ☐
(c) Design pressure ☐
(d) Hydraulic test pressure ☐

Q6. Section 4.3.1: responsibilities: definitions

Who does API 570 consider as ultimately responsible for developing, documenting, implementing, executing and assessing piping inspection systems and inspection procedures that will meet the requirements of API 570?

(a) The API 570 certified piping inspector ☐
(b) The plant user/owner themselves ☐
(c) An external QA audit body ☐
(d) All the above share equal responsibility ☐

Q7. Section 5.3: inspection for corrosion and cracking

How many areas/degradation mechanisms does API 570 specifically include in the 'at risk' list given in section 5.3?

(a) 7 ☐
(b) 8 ☐
(c) 10 ☐
(d) 12 ☐

Q8. Section 6.2: piping service classes

How many piping service classes are recognized by API 570 in its inspection periodicity table 6-1?

(a) 2 ☐
(b) 3 ☐
(c) 4 ☐
(d) 5 ☐

Q9. Section 7.1: corrosion rate determination

API (and ASME) place great importance on the effects of wall thinning of pipes/vessels and the calculation of the maximum allowable working pressure (i.e. design pressure) that a corroded item will stand. Table 7-1 of API 570 is based on which concept?

(a) The corrosion full-life concept ☐
(b) The corrosion half-life concept ☐

(c) The corrosion third-life concept ☐
(d) The corrosion quarter-life concept ☐

Q10. API 570 section 2: references

Which two of the following code combinations contain significant details on valves?

(a) API 570 and 571 ☐
(b) API 571 and 574 ☐
(c) API 574 and ASME B16.34 ☐
(d) API 564 and ASME IX ☐

Q11. Section 5.3.3.1: CUI

Within which temperature range is plain carbon steel piping particularly susceptible to CUI?

(a) −4 to 50 °C ☐
(b) −4 to 120 °C ☐
(c) 32 to 500 °F ☐
(d) 150 to 400 °F ☐

Q12. Section 8.1.1: authorization of piping repairs

When does the API-authorized piping inspector have to authorize piping repair work before it commences?

(a) Always ☐
(b) Only when the work is to be carried out by an
unauthorized piping repair company ☐
(c) Only when the repair work occurs outside the scope of
ASME B31.3 ☐
(d) Never, a qualified piping engineer can do it instead ☐

Q13. Section 6.3: inspection intervals

The inspection intervals given in API 570 table 6-1 are:

(a) Mandatory (in the USA) and cannot be changed ☐
(b) Only applicable to unlagged piping ☐
(c) Guidelines, which may be increased at the inspector's
discretion ☐
(d) Recommended maximum intervals which may be
decreased ☐

Q14. Section 8.3: re-rating, repair, alteration

A piping system is having one of its bends replaced with an identical item, owing to corrosion. At the same time the MAWP of the entire system is being increased owing to changed process conditions. Is this situation classed as?

(a) A repair only ☐
(b) A repair and an alteration ☐
(c) An alteration and a re-rating ☐
(d) A repair and a re-rating ☐

Q15. Appendix A: inspector recertification

According to API 570 appendix A, how often does an authorized API 570 inspector have to sit the on-line code-update test?

(a) Every year ☐
(b) Every three years ☐
(c) Every five years ☐
(d) Every six years ☐

Chapter 4

API 574

4.1 Introduction

This chapter of the book is about learning to become familiar with the layout and contents of API 574. It is a code that is strongly linked to API 570 and, in some areas, contains some of the same information that appears in API 570. We saw in the previous chapter how this linking of codes is a feature of the API/ASME approach to plant inspection (and the API certification exams).

API 574 is divided into twelve sections (sections 1 to 12) and one single-page appendix. The body of the text (i.e. sections 1 to 12) contains a large number of tables and figures, mainly sketches of piping components, interspersed with a few tables about pipe schedule wall thicknesses, etc.

The API 570 examination body of knowledge requires candidates to have knowledge of effectively *all* of the twelve sections. Hence API 574 is seen as a 'general piping knowledge' part of the API 570 certification syllabus. Fortunately, understanding the twelve sections is not as difficult as it first appears. In layout, API 574 is a heavily unbalanced code; i.e. a few sections (particularly section 10 covering inspection procedures) are quite long, up to fifteen pages, while most of them (sections 1, 2, 3, 5, 7, 8, 9, 11 and 12) are each only about one page long. In practice, sections 1, 2 and 3 are little more than preliminaries to the code and don't contain much new information.

The only way to approach API 574 is from a section-by-section viewpoint, to build a general understanding of its content. The best way of doing this is by reading important sections of the document and then attempting questions on the content.

4.2 API 574 section 4: piping components

Here are some of the important parts of API 574 section 4:

- **How pipe sizes and wall thicknesses are defined**. In section 4.1, look at table 1, table 1-A (for stainless steel pipe) and table 3 on tolerances (situated near the back of the code).
- **The different types of valves** in section 4.3. You need to be able to recognize them, not draw them. There are no drawing questions in the API exam.
- **The flanges and fittings** in figures 9 and 10. These are important as they are covered by many of the data tables in ASME B16.5.

4.3 API 574 familiarization questions

Look up the answers to these questions to help you become familiar with API 574.

Q1. API 574 section 4.1.1: general

Pipe wall thicknesses are designated as pipe *schedules* in what sizes of pipe?

(a) Up to 18 in ☐
(b) Up to 24 in ☐
(c) Up to 36 in ☐
(d) Up to 48 in ☐

Q2. API 574 section 4.1.1: general

What is an alternative system for defining pipe wall thicknesses?

(a) Standard, extra strong and double extra strong ☐
(b) Normal, strong and extra strong ☐
(c) Standard, strong and extra strong ☐
(d) XX, XXX and XXXX ☐

Q3. API 574 section 4.1.1: general

For NPS 12 pipe, which dimension stays the same, regardless of the wall schedule thickness?

(a) The inside diameter (ID) ☐
(b) The outside diameter (OD) ☐
(c) The mean wall diameter ☐
(d) None of these; they all vary with schedule thickness ☐

Q4. API 574 section 4.3.1: valves

Which ASME standard covers the pressure/temperature ratings of valve bodies?

(a) ASME B31.3 ☐
(b) ASME B16.5 ☐
(c) ASME B16.34 ☐
(d) ASME VIII ☐

Q5. API 574 table 3: ferritic pipe tolerances

What is the normal acceptable thickness under-tolerance on wall thickness for A106 plain carbon steel pipe up to NPS 48?

(a) −0.125 in ☐
(b) −10 % of nominal wall thickness ☐
(c) −12.5 % of nominal wall thickness ☐
(d) −15 % of nominal wall thickness ☐

4.4   Corrosion monitoring and inspection

Read through API 574 section 6, paying particular attention to the following topics:

- **The use of corrosion circuits** to help manage the inspections, calculations and record keeping relating to piping inspection. Section 6.2.1 gives various parameters that can be considered when identifying corrosion circuits.
- **The 14 areas/locations of degradation** listed in section 6.3. Note how twelve of them are repeats from a similar list in API 570.
- **The content of figure 21** in the code which shows TMLs marked up on a piping circuit diagram and section 6.3.1 about the inspection of injection points.

Then look at these important topics in API 574 section 10:

- **Thickness measurements: section 10.1.2.** Concentrate on the explanations of the limitations of UT measurement techniques
- **Radiographic inspection: section 10.1.2.2.** This gives a brief summary of the types of RT techniques that can be useful in piping inspection.

- **Pressure tests: section 10.2.3.** This explains the reason for pressure testing and some of its limitations.

4.5 API 574 (sections 6 and 10) familiarization questions

Try these questions to help you become familiar with these sections

Q1. API 574 section 10.1.2: corrosion monitoring of process piping

The API 574 view on what constitutes a good pipework monitoring programme is based heavily on?

(a) Monitoring pipework wall thickness ☐
(b) Monitoring fluid velocities to prevent erosion ☐
(c) Monitoring vibration using sensors ☐
(d) All of the above, as they share equal responsibility for pipe failures ☐

Q2. API 574 section 6.2.1: piping circuits

A piping 'corrosion circuit' comprises sections of piping of similar design and which:

(a) Are at the same temperature ☐
(b) Are at the same temperature and pressure ☐
(c) Are exposed to conditions of similar corrosivity ☐
(d) Appear on the same process and instrumentation diagram (P&ID) ☐

Q3. API 574 section 6.3: inspection for specific types of corrosion and cracking

Which two damage mechanisms are listed in section 6.3 of API 574 but do not appear in the similar table 5-3 in API 570?

(a) Injection points and deadlegs ☐
(b) CUI and soil–air interfaces ☐
(c) Fatigue cracking and creep cracking ☐
(d) Corrosion at pipe supports and dew-point corrosion ☐

Q4. API 574 section 6.3.1: injection points

When designating an injection point circuit (IPC) for the purposes of inspection, the recommended downstream limit of the IPC is:

(a) The first change in flow direction past the injection point ☐

(b) The second change in flow direction past the injection point ☐

(c) The third change in flow direction past the injection point ☐

(d) 25 feet past the second change in flow direction past the injection point ☐

Q5. API 574 section 6.3.1: injection points

The preferred method of inspecting injection points is?

(a) Dye penetrant and/or magnetic particle testing ☐

(b) Dye penetrant and/or ultrasonic testing ☐

(c) Ultrasonic testing ☐

(d) Radiographic testing and hammer testing ☐

Q6. API 574 section 10.2.3: pressure testing

As well as API 574, guidelines on pressure testing are found in?

(a) API 570 only ☐

(b) ASME B31.3 only ☐

(c) ASME B16.5 only ☐

(d) All of the above ☐

Q7. API 574 section 10.2.3: pressure testing

What is the main reason why it is necessary to bleed all of the air out before a hydraulic pressure test of a pipework system?

(a) Air absorbs water and so reduces the test pressure ☐

(b) For safety reasons ☐

(c) Air causes shock loadings and so an unsteady pressure gauge reading ☐

(d) Air heats up as it is compressed, further increasing the pressure ☐

Q8. API 574 section 10.2.3: pressure testing
Which material would not be suitable for testing with water containing chlorides (salts)?

(a) Ferritic plain carbon steels ☐
(b) High carbon steels ☐
(c) Austenitic stainless steels ☐
(d) Low-pressure steam pipes ☐

Q9. API 574 section 10.2.3: pressure testing
Which particular failure mechanism must be guarded against when doing a 'full pneumatic test'?

(a) Hydrogen failure ☐
(b) Creep rupture ☐
(c) Ductile fracture ☐
(d) Brittle fracture ☐

Q10. API 574 section 10.2.3: pressure testing
A test at low pressure (10 % MAWP) using air or nitrogen and a soap solution is known as?

(a) A hydraulic test ☐
(b) A hydrostatic test ☐
(c) A leak test ☐
(d) A full pneumatic test ☐

Chapter 5

API 578

5.1 Introduction

This short chapter covers the contents of API 578: *Material verification programme for new and existing alloy piping systems: 1999*. API 578 is a short document that has only recently been added to the syllabus for the API 570 and 510 ICP examinations.

API 578 is divided into seven sections (sections 1 to 7). These are all descriptive sections (no calculations) that cover the various aspects of operating what API call a *Material Verification Programme*. In Europe this is more commonly known as Positive Material Identification (PMI).

The API 570 syllabus describes the knowledge required in ten areas, which (broadly) mirror the API 578 sections. This means that, unlike for some of the codes, all of the sections of API 578 are covered in the syllabus.

One important feature of API 578 is that it is very much a 'stand-alone' code with few cross-references to the other API codes. Historically, ASME and API codes for pressure equipment (and many other things as well) have made little reference to material traceability. In Europe the DIN 50049/ EN10204 standard has for many years set requirements for material certification, but such an approach has not been followed as enthusiastically in the US. There is an ASTM standard on material traceability but it does not have particularly wide use.

API 578 is therefore useful in filling a gap in the scope of US codes. Note, however, that it adopts a different approach from EN 10204. API 578 concentrates on *retrospective* PMI rather than 'at source' material/mill certificates.

In essence, API 578 was introduced to fill a perceived gap in the coverage of other API documents. In practice, it can be applied to both new construction and in-service systems, including after repairs have been carried out. Note the following points:

- One of the main objectives of material verification is to avoid mix-ups between low carbon steels and low-alloy steel containing Cr, Mn and Mo. Physically these materials look very similar (unlike stainless steel which looks different) so mix-ups are perhaps more likely.
- The amount of verification needed depends on the level of risk that exists; i.e. how good was the original material specification and control and/or could substitution of incorrect material result in additional corrosion or risk of failure?

5.2   The verification/PMI techniques

API 578 sets out several techniques that can be used. The main ones are:

- **Portable X-ray fluorescence (XRF)**. In this technique, gamma rays excite the material, causing it to emit X-rays. These are analysed, indicating the elements present in the material.
- **Emission spectrometry**. Here, an electric arc stimulates the atoms in the material. The emitted spectrum shows which atoms are present.

5.3   API 578 familiarization questions

Try these questions to help you become familiar with API 578.

Q1.   API 578 section 1.3: documentation of roles and responsibilities

An important requirement of a Material Verification Programme is that:

(a) Roles and responsibilities should be in accordance with
      API 570   ☐
(b) Roles and responsibilities should be clearly documented   ☐
(c) A Material Verification Programme only involves a
      single party   ☐

(d) A Material Verification Programme is always written by the user ☐

Q2. API 578 section 3.1: definitions
What is the API 578 definition of an 'alloy material'?

(a) A material containing > 0.3 % carbon ☐
(b) A material known as 'stainless steel' containing > 12 % Cr ☐
(c) A metallic material containing alloying elements such as Cr, Ni or Mo ☐
(d) Any material containing alloying elements such as Cr, Ni or Mo ☐

Q3. API 578 section 4.1: extent of PMI verification
For high-risk piping systems where material mix-ups are considered likely, API 578 suggests an examination of what percentage of the material should be considered:

(a) 10 %. ☐
(b) 50 % ☐
(c) 100 % ☐
(d) No percentage figure is given in API 578 ☐

Q4. API 578 section 4.2.1: responsibilities
Who has the *ultimate* responsibility for deciding the extent of PMI required and to verify that the PMI programme is properly performed?

(a) The owner/user ☐
(b) The material manufacturer or welding contractor (for repairs) ☐
(c) A third-party inspector ☐
(d) An API-certified inspector ☐

Q5. API 578 section 4.2.6: PMI testing
How does API 578 suggest that weld consumables should be tested as part of a Material Verification Programme?

(a) There is no need to test consumables that have certificates with them ☐
(b) One electrode or wire sample per lot (batch) should be tested ☐
(c) 10 % of consumables per lot (batch) should be tested ☐
(d) Testing should only be done after the welding is completed ☐

Q6. API 578 section 4.2.7: PMI of bought-out components

For components that are bought-in from a distributor (stockist) API 578 recommends that it is advisable for the plant user (i.e. the purchaser) to:

(a) Do *less* PMI, as it is mainly the responsibility of the
    stockist ☐
(b) Do *more* PMI, as the risk of mix-ups is higher ☐
(c) Ignore PMI and rely on the certificate of conformity
    from the stockist ☐
(d) Do PMI of 10 % of the components ☐

Q7. API 578 section 4.3.2.1: material mix-ups

Which of these historically results in the largest chance of mix-ups between the types of material?

(a) Low carbon steel and high carbon steel ☐
(b) Low carbon steel and stainless steel ☐
(c) Low carbon steel and low alloy steel ☐
(d) Stainless steel and low alloy steel ☐

Q8. API 578 section 4.3.3: material mix-ups

Which of these components have a *lower-than-average* risk of material mix-ups?

(a) Small diameter piping (< 2 in diameter) ☐
(b) Welding consumables (electrodes or wire) ☐
(c) Bolts ☐
(d) Valves ☐

Q9. API 578 section 5: PMI techniques

Which of these is *not* a PMI technique mentioned in API 578?

(a) XRF ☐
(b) Emission spectrometry ☐
(c) Resistivity testing ☐
(d) Replica testing ☐

Q10. API 578 section 6.2: dissimilar welds

What is a potential problem in doing XRF or similar PMI testing of a weld containing dissimilar materials?

(a) Dilution ☐
(b) The Seebeck effect ☐
(c) The equipment is difficult to calibrate for dissimilar welds ☐
(d) The resistivity effect distorts the results ☐

Chapter 6

API 571

6.1   API 571: Introduction

This chapter covers the contents of API 571: *Damage mechanisms affecting fixed equipment in the refining industry: 2003*. API 571 has only recently been added to the syllabus for the API 570 and 510 examinations and replaces what used to be included in an old group of documents dating from the 1960s entitled IRE (*Inspection of refinery equipment*).

The first point to note is that the API 571 sections covered in the API 570 ICP exam syllabus is only an extract from the full version of API 571.

6.1.1   The fifteen damage mechanisms

Your copy of API 571 contains (among other things) descriptions of fifteen damage mechanisms (we will refer to them as DMs). Here they are in Fig. 6.1.

Remember that these are all DMs that are found in the

1. Brittle fracture
2. Thermal fatigue
3. Erosion–corrosion
4. Mechanical fatigue
5. Vibration-induced fatigue
6. Atmospheric corrosion
7. Corrosion under insulation (CUI)
8. Boiler condensate corrosion
9. Flue-gas dewpoint corrosion
10. Microbiological-induced corrosion (MIC)
11. Soil corrosion
12. Sulfidation
13. SCC
14. Caustic embrittlement
15. High-temperature hydrogen attack (HTHA)

Figure 6.1   Fifteen API 571 damage mechanisms

petrochemical/refining industry (because that is what API 571 is about), but they may or may not be found in other industries. Some, such as brittle fracture and fatigue, are commonly found in non-refinery plant whearas others, such as sulfidation, are not.

6.1.2 Are these DMs in some kind of precise logical order?

No, or if they are, it is difficult to see what it is. The list contains a mixture of high- and low-temperature DMs, some of which affect plain carbon steels more than alloy or stainless steels and vice versa. There are also various subdivisions and a bit of repetition thrown in for good measure. None of this is worth worrying about, as the order in which they appear is not important.

In order to make the DMs easier to remember you can think of them as being separated into three groups. There is no code significance in this rearrangement at all; it is simply to make them easier to remember. Figure 6.2 shows the

• Brittle fracture • Thermal fatigue • Mechanical fatigue • Vibration-induced fatigue	**Group 1** **Common DMs**
• Erosion corrosion • Atmospheric corrosion • Corrosion under insulation (CUI) • Soil corrosion	**Group 2** **Environment-related** **(more or less)**
• Boiler condensate corrosion • Flue-gas dewpoint corrosion • Microbiological-induced corrosion (MIC) • Sulfidation • SCC • Caustic embrittlement • High-temperature hydrogen attack (HTHA)	**Group 3** **Higher** **temperature/more** **refinery-specific**

Figure 6.2   DMs: a revised order

revised order in which they appear in the familiarization questions.

One important feature of API 571 is that it describes each DM in some detail, with the text for each one subdivided into *six subsections*. Figure 6.3 shows the subsections and the order in which they appear.

These six subsections are *important* as they form the subject matter from which the API examination questions are taken. As there are no calculations in API 571, and only a few graphs, etc., of detailed information, you can expect most of the API examination questions to be *closed book*, i.e.

REMEMBER THE WAY THAT API 571 COVERS EACH OF THE DAMAGE MECHANISMS

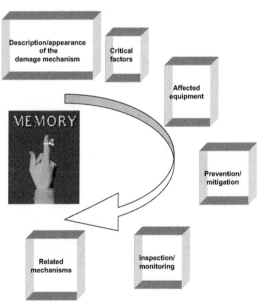

Figure 6.3   The content 'subsections' of API 571

a test of your understanding and short-term memory of the DMs. The questions could come from any of the six subsections shown in Fig. 6.3.

6.2 The first group of DMs

The following Figs 6.4 to 6.7 relate to the first group of DMs in API 571. When looking through these figures, try to cross-reference them to the content of the relevant sections of API 571. Then read the full sections of API 571 covering the four DMs in this first group.

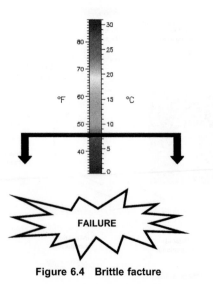

Caused by hydro-testing and/or operating below the Charpy impact transition temperature

Figure 6.4   Brittle facture

On a macro scale, thermal fatigue cracks tend to be dagger-shaped, wide and oxide-filled (caused by the oxidizing effect of the temperature variations)

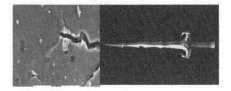

Joint restraint is a common cause of excessive thermal stresses

Figure 6.5   Thermal fatigue

The result of cyclic stresses caused by mechanical loadings

Common around stress concentrations
- Changes of section
- Keyways
- Rough welds
- Thread notches

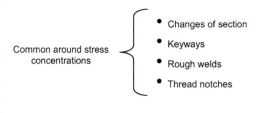

Fatigue crack in progress   Propagation to failure

Figure 6.6   Mechanical fatigue

This is simply mechanical fatigue caused by induced vibrations

Typical causes:

- Water hammer
- 'Flash-off' of fluid
- Small-bore pipes that are unsupported
- Vortex-vibration in heat exchangers
- Failure of pipe hangers

Failures commonly occur at socket welded or threaded joints: like this

Figure 6.7 Vibration-induced fatigue

6.3 API 571 familiarization questions (set 1)

Attempt this first set of self-test questions covering the first group of API 571 DMs.

Q1. API 571 section 4.2.7.1: brittle fracture

Which of these is a description of brittle fracture?

(a) Sudden rapid fracture of a material with plastic deformation ☐
(b) Sudden rapid fracture of a material without plastic deformation ☐
(c) Unexpected failure as a result of cyclic stress ☐
(d) Fracture caused by reaction with sulfur compounds ☐

Q2. API 571 section 4.2.7.2: brittle fracture: affected materials

Which of these materials are particularly susceptible to brittle fracture?

(a) Plain carbon and high alloy steels ☐
(b) Plain carbon, low alloy and 300 series stainless steels ☐
(c) Plain carbon, low alloy and 400 series stainless steels ☐
(d) High-temperature resistant steels ☐

Q3. API 571 section 4.2.7.3: brittle fracture: critical factors

At what temperature is brittle fracture most likely to occur?

(a) Temperatures above 400 °C ☐
(b) Temperatures above the Charpy impact transition temperature ☐
(c) Temperatures below the Charpy impact transition temperature ☐
(d) In the range 20–110 °C ☐

Q4. API 571 section 4.2.7.4: brittle fracture

Which of these activities is *unlikely* to result in a high risk of brittle fracture?

(a) Repeated hydrotesting above the Charpy impact transition temperature ☐
(b) Initial hydrotesting at low ambient temperatures ☐
(c) Start-up of thick-walled vessels ☐
(d) Autorefrigeration events ☐

Q5. API 571 section 4.2.7.6: brittle fracture: prevention/mitigation

What type of material change will *reduce* the risk of brittle fracture?

(a) Use a material with lower toughness ☐
(b) Use a material with lower impact strength ☐
(c) Use a material with a higher ductility ☐
(d) Use a thicker material section ☐

Q6. API 571 section 4.2.7.5: brittle fracture: appearance

Cracks resulting from brittle fracture will most likely be predominantly:

(a) Branched ☐
(b) Straight and non-branching ☐
(c) Intergranular ☐
(d) Accompanied by localized necking around the crack ☐

Q7. API 571 section 4.2.9: thermal fatigue: description

What is thermal fatigue?

(a) The result of excessive temperatures ☐
(b) The result of temperature-induced corrosion ☐

(c) The result of cyclic stresses caused by temperature variations ☐
(d) The result of cyclic stresses caused by dynamic loadings ☐

Q8. API 571 section 4.2.9.3: thermal fatigue: critical factors

As a practical rule, thermal cracking may be caused by temperature swings of *approximately*:

(a) 200 °C ☐
(b) 200 °F ☐
(c) 100 °C ☐
(d) 100 °F ☐

Q9. API 571 section 4.2.9.5: thermal fatigue: appearance

Cracks resulting from thermal fatigue will most likely be predominantly:

(a) Straight and non-branching ☐
(b) Dagger-shaped ☐
(c) Intergranular ☐
(d) Straight and narrow ☐

Q10. API 571 section 4.2.9.6: prevention/mitigation

Thermal fatigue cracking is best avoided by:

(a) Better material selection ☐
(b) Control of design and operation ☐
(c) Better post-weld heat treatment (PWHT) ☐
(d) Reducing mechanical vibrations ☐

6.4 The second group of DMs

Figures 6.8 to 6.10 relate to the second group of DMs. Note how these DMs tend to be environment-related. Remember to identify the six separate subsections in the text for each DM.

High fluid velocities cause scouring. Bends and welds are particularly susceptible

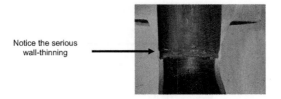

Notice the serious wall-thinning →

TO REDUCE EROSION–CORROSION

- Reduce fluid velocity
- Use a more resistant material (harder alloys may be better)

Figure 6.8   Erosion/corrosion

CUI hides under lagging, and is often widespread

Chloride contamination (from water or lagging) makes CUI much worse

Figure 6.9   Corrosion under insulation (CUI)

Affects pipework/structures *on* or *under* the soil. Causes heavy pitting and wall thinning

Soil/air interfaces are particularly susceptible to corrosion

Figure 6.10   Soil corrosion

6.5   API 571 familiarization questions (set 2)

Q1.   API 571 section 4.2.14
A damage mechanism that is strongly influenced by fluid velocity and the corrosivity of the process fluid is known as:

(a) Mechanical fatigue ☐
(b) Erosion–corrosion ☐
(c) Dewpoint corrosion ☐
(d) Boiler condensate corrosion ☐

Q2.   API 571 section 4.3.2: atmospheric corrosion
As a practical rule, atmospheric corrosion:

(a) Only occurs under insulation ☐
(b) May be localized or general (widespread) ☐
(c) Is generally localized ☐
(d) Is generally widespread ☐

Q3.   API 571 section 4.3.2.3: atmospheric corrosion: critical factors
A typical atmospheric corrosion rate in mils (1 mil = 0.001 in) per year (mpy) of steel in an inland location with moderate precipitation and humidity is:

(a) 1–3 mpy ☐
(b) 5–10 mpy ☐
(c) 10–20 mpy ☐
(d) 50–100 mpy ☐

Q4. API 571 section 4.3.3.3: CUI critical factors
Which of these metal temperature ranges will result in the most severe CUI?

(a) 0–51 °C ☐
(b) 100–121 °C ☐
(c) 0 to −10 °C ☐
(d) 250+ °C ☐

Q5. API 571 section 4.3.3.6: CUI appearance
Which other corrosion mechanism often accompanies CUI in 300 series stainless steels?

(a) HTHA ☐
(b) Erosion–corrosion ☐
(c) Dewpoint corrosion ☐
(d) SCC ☐

Q6. API 571 section 4.3.3.6: CUI prevention/mitigation
Which of these would *significantly reduce the risk* of the occurrence of CUI on a 316 stainless steel pipework system fitted with standard mineral wool lagging?

(a) Change the pipe material to a 304 stainless steel ☐
(b) Change to a calcium silicate lagging material ☐
(c) Shotblast the pipe surface and re-lag ☐
(d) Change to low-chloride lagging material ☐

Q7. API 571 section 4.3.9: soil corrosion
What is the main parameter measured to assess the corrosivity of a soil?

(a) Acidity ☐
(b) Alkalinity ☐
(c) Density ☐
(d) Resistivity ☐

Q8. API 571 section 4.3.9.5: soil corrosion: appearance
What would you expect the result of soil corrosion to look like?

(a) Branched and dagger-shaped cracks ☐
(b) Isolated, large and deep individual pits ☐
(c) Straight cracks ☐
(d) External corrosion with wall thinning and pitting ☐

Q9. API 571 section 4.3.9.6: soil corrosion protection
Which of these would be used to *reduce* the amount of soil corrosion?

(a) Caustic protection ☐
(b) Cathodic protection ☐
(c) Anodic protection ☐
(d) More post-weld heat treatment ☐

Q10. API 571 section 4.3.9.3: soil corrosion critical factors
What effect does metal temperature have on the rate of soil corrosion?

(a) None ☐
(b) The corrosion rate increases with temperature ☐
(c) The corrosion rate decreases with temperature ☐
(d) There will be minimum soil corrosion below 0 °C ☐

6.6 The third group of DMs

Now look through Figs 6.11 to 6.16 covering the final group of seven DMs. These DMs tend to be either more common at higher temperatures or a little more specific to refinery equipment than those in the previous two groups.

Again, remember to identify the six separate subsections in the text for each DM, trying to anticipate the type of examination questions that could result from the content.

pH and temperature make it worse

Extreme cases can lead to leakage

Figure 6.11 Oxygen pitting

Caused by living microbial organisms

- Bacteria
- Algae
- Fungi

Normally found where aqueous conditions (water) are present

Causes localized pitting (sometimes under tubercle 'caps')

Figure 6.12   Microbial-induced corrosion (MIC)

This is a high-temperature corrosion mechanism

Sulfur compounds

High temperatures
(260°C +)   } Sulfidation corrosion

Carbon and alloy steels

The main problem is caused by H_2S (formed by the degradation of sulfur
compounds at high temperature)

Occurs in crude plant, cokers, hydroprocessor units, fired heaters, etc. –
anywhere where there are high-temperature sulfur streams

Figure 6.13   Sulfidation corrosion

API terminology calls it 'Environmental-*assisted cracking*'

The stress exposes the grain boundaries to corrosion

Temperature range above 60°C (140 °F) and pH > 2

- One of the most common corrosion mechanisms

- Prevalent in 300 series austenitic stainless steel and high-chromium alloys

Figure 6.14   Stress corrosion cracking (SCC)

A specialist type of stress corrosion cracking caused by alkaline conditions. The worst offenders are:

- Caustic potash (KOH)

- Sodium hydroxide (NaOH)

Caustic attack in a heat exchanger tubesheet

Typically found in H_2S removal units and acid neutralization units

Figure 6.15   Caustic embrittlement

This is a specialist and complex corrosion mechanism

At high temperatures, H_2 reacts with the carbon in the steel forming CH_4 (methane)

The resulting loss of carbides weakens the steel

Fissures start to form and propagate into cracks

Figure 6.16   High-temperature hydrogen attack (HTHA)

6.7 API 571 familiarization questions (set 3)

Now try the final set of API 571 self-test questions, covering the third group of DMs.

Q1. API 571 section 4.3.5: boiler water condensate corrosion

Which substance is the main cause of corrosion (common to several corrosion mechanisms) in a boiler water condensate system?

(a) Hydrogen ☐
(b) Chlorine ☐
(c) Hydrazine ☐
(d) Oxygen ☐

Q2. API 571 section 4.3.5.5: boiler water condensate corrosion: appearance

What does CO_2 corrosion of pipe internals look like?

(a) Smooth grooving of the pipe wall ☐
(b) Blisters (tubercles) on the pipe wall ☐
(c) Branched cracks in the pipe wall ☐
(d) Cup-shaped pits and a slimy surface on the pipe wall ☐

Q3. API 571 section 4.3.7: dew-point corrosion

Dew-point corrosion in flue gas systems in caused by:

(a) The presence of living organisms ☐
(b) High temperatures ☐
(c) Sulfur and chlorine compounds ☐
(d) Ash, particulates and other non-combustibles ☐

Q4. API 571 section 4.3.7.3: dew-point corrosion: critical factors

Dew-point corrosion in a system where the flue gas contains SO_2 and SO_3 occurs at what metal temperatures?

(a) Below 138 °C ☐
(b) Any temperature above 138 °C ☐
(c) 138–280 °C ☐
(d) Only below 54 °C ☐

Q5. API 571 section 4.3.8: microbiologically induced corrosion (MIC)
What does MIC typically look like?

(a) Uniform wall thinning with a 'sparkling' corroded surface ☐
(b) Localized pitting under deposits ☐
(c) Smooth longitudinal grooving ☐
(d) A dry flaky appearance ☐

Q6. API 571 section 4.4.2: sulfidation: critical factors
Sulfidation is the reaction of steels with sulfur compounds at what temperatures?

(a) Any temperature ☐
(b) 54–260 °C ☐
(c) Above 260 °C ☐
(d) Above 380 °C ☐

Q7. API 571 section 4.4.2.4: sulfidation: affected equipment
Which of these items of equipment would be most likely to suffer from sulfidation?

(a) Natural gas-fired power boilers ☐
(b) Residual oil-fired boilers ☐
(c) High-pressure compressed air systems ☐
(d) Sea water cooling systems ☐

Q8. API 571 section 4.4.2: sulfidation
Which curves document the effect of temperature on sulfidation rates in steels?

(a) The NACE economy curves ☐
(b) The McConomy curves ☐
(c) The Larsen–Miller curves ☐
(d) The morphology curves ☐

Q9. API 571 section 4.4.2: sulfidation
Which of these materials has the highest resistance to sulfidation?

(a) Low carbon steel ☐
(b) $2\frac{1}{4}$ % Cr alloy steel ☐
(c) 9 % Cr alloy steel ☐
(d) 18/8 stainless steel ☐

Q10. API 571 section 4.5: environment-assisted cracking

According to API, what are the two main types of environment-assisted cracking?

(a) Chloride SCC and CUI ☐
(b) Chloride SCC and sulfur SCC ☐
(c) Chloride SCC and hydrogen attack ☐
(d) Chloride SCC and caustic embrittlement ☐

SECTION II: WELDING

Chapter 7

Introduction to Welding/API 577

7.1 Introduction

The purpose of this chapter is to ensure you can recognize the main welding processes that may be specified by the welding documentation requirements of ASME IX. The API exam will include questions in which you have to assess a Weld Procedure Specification (WPS) and its corresponding Procedure Qualification Record (PQR). As the codes used for API certification are all American you need to get into the habit of using American terminology for the welding processes and the process parameters.

This module will also introduce you to the API RP 577: *Welding inspection and metallurgy* in your code document package. This document has only recently been added to the API examination syllabus. As a Recommended Practice (RP) document, it contains technical descriptions and instruction, rather than truly prescriptive requirements.

7.2 Welding processes

There are four main welding processes that you have to learn about:

- Shielded metal arc welding (SMAW)
- Gas tungsten arc welding (GTAW)
- Gas metal arc welding (GMAW)
- Submerged arc welding (SAW)

The process(es) that will form the basis of the WPS and PQR questions in the API exam will almost certainly be chosen from these.

The sample WPS and PQR forms given in the non-mandatory appendix B of ASME IX (the form layout is not strictly within the API 570 examination syllabus, but we will discuss it later) *only* contains the information for qualifying these processes.

7.2.1 Shielded metal arc (SMAW)

This is the most commonly used technique. There is a wide choice of electrodes, metal and fluxes, allowing application to different welding conditions. The gas shield is evolved from the flux, preventing oxidation of the molten metal pool (Fig. 7.1). An electric arc is then struck between a coated electrode and the workpiece. SMAW is a manual process as the electrode voltage and travel speed is controlled by the welder. It has a constant current characteristic.

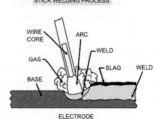

- An electric arc is struck between a consumable flux-coated wire electrode and the workpiece

- It is a manual process because the welding electrode voltage and travel speed are controlled by the welder

- It has a constant current characteristic

Commonly known in Europe as

- MMA – manual metal arc welding

or

- 'Stick' welding

Figure 7.1   SMAW process

7.2.2 Metal inert gas (GMAW)

In this process, electrode metal is fused directly into the molten pool. The electrode is therefore consumed, being fed from a motorized reel down the centre of the welding torch (Fig. 7.2). GMAW is known as a semi-automatic process as the welding electrode voltage is controlled by the machine.

Tungsten inert gas (GTAW)

This uses a similar inert gas shield to GMAW but the tungsten electrode is not consumed. Filler metal is provided from a separate rod fed automatically into the molten pool (Fig. 7.3). GTAW is another manual process as the welding electrode voltage and travel speed are controlled by the welder.

Submerged arc welding (SAW)

In SAW, instead of using shielding gas, the arc and weld zone are completely submerged under a blanket of granulated flux (Fig. 7.4). A continuous wire electrode is fed into the weld.

 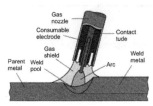

- An electric arc is struck between a continuously fed consumable solid electrode wire and the workpiece

- It is known as a semi-automatic process because the welding electrode voltage is controlled by the machine

Also known in Europe as

MIG – metal inert gas welding

or

MAG – metal active gas welding

Figure 7.2   GMAW process

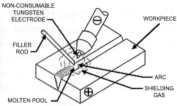

- An electric arc is struck between a non-consumable tungsten electrode and the workpiece. Filler rod is added separately

- It is a manual process when the welding electrode voltage and travel speed are controlled by the welder

Also known in Europe as

TIG – tungsten inert gas welding

or (rarely)

TAG – tungsten active gas welding

Figure 7.3   GTAW process

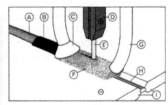
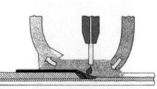

Schematic representation of submerged arc welding

A Finished weld   F Flux
B Slag   G Flux supply
C Powder removal   H Root bead
D Electrode holder   I Parent metal
E Filler wire

- An electric arc is struck between a reel-fed continuous consumable electrode wire and the work with the arc protected underneath a flux blanket

- It can be a semi-automatic, mechanized or automated process

Figure 7.4   SAW process

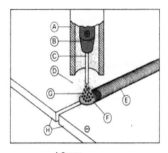

In FCAW the filler wire contains a flux. This protects the weld from the atmosphere by coating it in a slag (similar to SAW)

A Gas cup
B Electrode holder
C Filler wire
D Shielding gas
E Finished weld
F Weld pool
G Arc
H Parent metal

Figure 7.5   FCAW process

This is a common process for welding structural carbon or carbon-manganese steelwork. It is usually automatic, with the welding head being mounted on a traversing machine. Long continuous welds are possible with this technique.

Flux-cored arc welding (FCAW)
FCAW is similar to the GMAW process, but uses a continuous hollow electrode filled with flux, which produces the shielding gas (Fig. 7.5). The advantage of the technique is that it can be used for outdoor welding, as the gas shield is less susceptible to draughts.

7.3   Welding consumables
An important area of the main welding processes is that of weld *consumables*. We can break these down into the following three main areas:

- Filler (wires, rods, flux-coated electrodes)
- Flux (granular fluxes)
- Gas (shielding, trailing or backing)

There are always questions in the API examination about weld consumables. Figures 7.6 to 7.11 show basic information about the main welding processes and their consumables.

WELDING CONSUMABLES

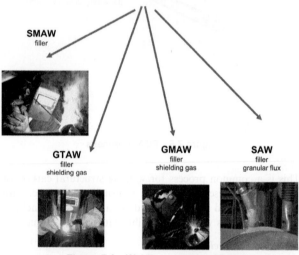

SMAW
filler

GTAW
filler
shielding gas

GMAW
filler
shielding gas

SAW
filler
granular flux

Figure 7.6   Welding consumables

TYPES

- Basic: for low-hydrogen applications
- Rutile: for general purpose applications
- Cellulosic: for stovepipe (vertical down) applications

FILLER
Flux-coated electrodes

Figure 7.7   SMAW consumables

The American Welding Society (AWS) have a welding electrode identification system (see API 577 section 7.4 and appendix A).This is the system for SMAW electrodes

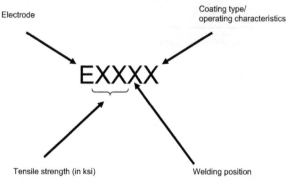

Electrode

Coating type/ operating characteristics

EXXXX

Tensile strength (in ksi)

Welding position

Figure 7.8   SMAW consumables identifications

FILLER:
Solid rods or wire

SHIELDING GAS:
Argon, helium or mixtures

Figure 7.9   GTAW consumables

FILLER:
Solid wire supplied on reels

SHIELDING GAS:
Inert gases – argon , helium, or mixtures. Active gases – carbon dioxide (CO_2) or Ar/CO_2 mixtures

Figure 7.10   GMAW consumables

FILLER:
Solid wire supplied on reels

FLUX:
Agglomerated: for low-hydrogen applications
Fused: for general applications

Figure 7.11   SAW consumables

7.4 Welding process familiarization questions

Q1.   API 577 section 5.2

How is fusion obtained using the SMAW process?

(a) An arc is struck between a consumable flux-coated
electrode and the work   ☐

(b) An arc is struck between a non-consumable electrode
and the work   ☐

(c) The work is bombarded with a stream of electrons and protons ☐
(d) An arc is struck between a reel-fed flux-coated electrode and the work ☐

Q2. API 577 section 5.1
Which of the following is not an arc welding process?

(a) SMAW ☐
(b) STAW ☐
(c) GMAW ☐
(d) GTAW ☐

Q3. API 577 section 5.3
How is fusion obtained using the GTAW process?

(a) An arc struck between a consumable flux-coated electrode and the work ☐
(b) An arc between a non-consumable tungsten electrode and the work ☐
(c) The work is bombarded with a stream of electrons and protons ☐
(d) An arc is struck between a reel-fed flux-coated electrode and the work ☐

Q4. API 577 section 5.3
How is the arc protected from contaminants in GTAW?

(a) By the use of a shielding gas ☐
(b) By the decomposition of a flux ☐
(c) The arc is covered beneath a fused or agglomerated flux blanket ☐
(d) All of the above methods can be used ☐

Q5. API 577 section 5.4
How is fusion obtained using the GMAW process?

(a) An arc struck between a consumable flux-coated electrode and the work ☐
(b) An arc between a non-consumable electrode and the work ☐
(c) The work is bombarded with a stream of electrons and protons ☐
(d) An arc is struck between a continuous consumable electrode and the work ☐

Q6. API 577 section 5.4

Which of the following are modes of metal transfer in GMAW?

(a) Globular transfer ☐
(b) Short circuiting transfer ☐
(c) Spray transfer ☐
(d) All of the above ☐

Q7. API 577 section 5.6

How is the arc shielded in the SAW process?

(a) By an inert shielding gas ☐
(b) By an active shielding gas ☐
(c) It is underneath a blanket of granulated flux ☐
(d) The welding is carried out underwater ☐

Q8. API 577 section 5.6

SAW stands for:

(a) Shielded Arc Welding ☐
(b) Stud Arc Welding ☐
(c) Submerged Arc Welding ☐
(d) Standard Arc Welding ☐

Q9. API 577 sections 5.3 and 3.7

Which of the following processes can weld *autogenously*?

(a) SMAW ☐
(b) GTAW ☐
(c) GMAW ☐
(d) SAW ☐

Q10. API 577 section 5.3.1

Which of the following is a commonly accepted advantage of the GTAW process?

(a) It has a high deposition rate ☐
(b) It has the best control of the weld pool of any of the arc processes ☐
(c) It is less sensitive to wind and draughts than other processes ☐
(d) It is very tolerant of contaminants on the filler or base metal ☐

7.5 Welding consumables familiarization questions

Q1

In a SMAW electrode classified as E7018 what does the 70 refer to?

(a) A tensile strength of 70 ksi ☐
(b) A yield strength of 70 ksi ☐
(c) A toughness of 70 J at 20 °C ☐
(d) None of the above ☐

Q2

Which of the following does not produce a layer of slag on the weld metal?

(a) SMAW ☐
(b) GTAW ☐
(c) SAW ☐
(d) FCAW ☐

Q3

Which processes use a shielding gas?

(a) SMAW and SAW ☐
(b) GMAW and GTAW ☐
(c) GMAW, SAW and GTAW ☐
(d) GTAW and SMAW ☐

Q4

What type of flux is used to weld a low-hydrogen application with SAW?

(a) Agglomerated ☐
(b) Fused ☐
(c) Rutile ☐
(d) Any of the above ☐

Q5

What shielding gases can be used in GTAW?

(a) Argon ☐
(b) CO_2 ☐
(c) Argon/CO_2 mixtures ☐
(d) All of the above ☐

Q6

Which process does not use bare wire electrodes?

(a) GTAW ☐
(b) SAW ☐
(c) GMAW ☐
(d) SMAW ☐

Q7

Which type of SMAW electrode would be used for low-hydrogen applications?

(a) Rutile ☐
(b) Cellulosic ☐
(c) Basic ☐
(d) Reduced hydrogen cellulosic ☐

Q8

In an E7018 electrode, what does the 1 refer to?

(a) Type of flux coating ☐
(b) It can be used with AC or DC ☐
(c) The positional capability ☐
(d) It is for use with DC only ☐

Q9

Which of the following processes requires filler rods to be added by hand?

(a) SMAW ☐
(b) GTAW ☐
(c) GMAW ☐
(d) SAW ☐

Q10

Which of the following process(es) use filler supplied on a reel?

(a) GTAW ☐
(b) SAW ☐
(c) GMAW ☐
(d) Both (b) and (c) ☐

Chapter 8

General Welding Rules of ASME B31.3 and API 570

8.1 Introduction

This chapter is to introduce you to the main general welding rules contained in API 570 and ASME B31.3. API 570 is for *in-service* inspection of process piping and therefore most welding carried out will be *repair* welds, rather than welds carried out on new systems. Much of API 570 is also concerned with the repair of corroded areas and on-stream systems (hot tapping) where process fluid is still being carried through the piping.

API 570 (section 8.2) makes it a *mandatory* requirement to comply with the welding rules contained in ASME B31.3 because it says that *all repair and alteration welding **shall** be done in accordance with the principles of ASME B31.3*. In practice, this will only be for areas that do not directly conflict with API 570 requirements.

ASME B31.3 lays down the rules for the design and construction of *new* process piping systems. The welding requirements are laid down in B31.3 chapter V. We will also look later at the inspection requirements (including acceptance criteria) that are laid down in B31.3 chapter VI. In essence, many areas of welding and inspection are common to both codes with only a few areas that differ between codes. This means that, in practice, most repair welding on pipework is done in accordance with the requirements of ASME B31.3.

8.2 Welding requirements of API 570

The following is a shortened version of the specific requirements specified in API 570 for carrying out repair and alteration welds. Read through this carefully. If you have any doubt about the meaning, look at the relevant clause in

API 570 to clarify it. The clause numbers match those in the code. Note that there are two sections to look at: section 8 and appendix C.

Section 8.1.1 authorization (what the inspector has to approve)

The inspector has to authorize any *repairs* prior to work starting. Note that he will only authorize alteration work after gaining approval from the piping engineer. As a general principle, the inspector may give prior authorization for fairly straightforward, limited or routine repairs to competent repair organizations, providing he is happy that they are competent.

Section 8.1.2: approval

There are four significant requirements here:

- The inspector (or piping engineer if more appropriate) must approve all aspects of the repair.
- The inspector can approve repairs at designated hold points.
- The piping engineer must be consulted before in-service cracks are repaired.
- Owner/user approval of on-stream welding is required.

Section 8.1.3.1: temporary repairs (note the important restrictions)

Temporary repairs consist of a full encirclement split sleeve or box-type enclosure for damaged or corroded areas (but not for longitudinal cracks, which could propagate under the sleeve).

Localized pitting or pinholes can be repaired with a fillet-welded split coupling or plate patch as long as:

- The specified minimum yield (SMYS) of the pipe is ≤ 40 000 psi (275 800 kPa).
- The coupling or patch is properly designed.
- The material matches the base material.

Enclosures (this means patches, sleeves, etc.) can be welded over minor leaks while the system is in service as long as the inspector is satisfied that further damage will not be caused.

Repairs should be replaced at next available maintenance opportunity unless an extension is approved and documented by the piping engineer.

Section 8.1.3.2: permanent repairs
Permanent repairs to pipework may consist of:

- preparing a groove to remove defects and then filling it with weld metal;
- restoring corroded areas by weld metal deposition;
- replacing sections of piping.

The other option is to use flush-inserts (rather than fillet welded patches on top of the corroded section). Note, however, the three requirements for this:

- Full penetration groove welds have to be used.
- Class 1 or 2 (this refers to API 570 fluid service classes: see API 570 section 6.2) system welds must be 100 % radiographed or ultrasonically tested.
- The inserts must have rounded corners with a minimum 1 in (25 mm) radius.

Section 8.2: welding and hot tapping
API 570 specifies that welding has to be done in accordance with ASME B31.3 for repairs and alterations.

Welding on systems *in operation* has to be in accordance with API publication 2201 and the inspector has to use the 'suggested hot tapping checklist' in API publication 2201 as a minimum. **Note**: don't get worried…this API 2201 document is not in the API 570 examination scope (you just need to know that it exists and what it refers to).

The reason for the difference between cold repair welding and hot tapping is that welds on systems in operation can be very dangerous due to the process fluid still being present. Preheating may be required due to the cooling effect the flowing process fluid has on the base material. There is also the risk of burning through the pipe and igniting the process fluid, with catastrophic results if it is a flammable gas or liquid.

Section 8.2.1: procedures, qualifications and records

This specifies that the organization doing weld repairs on pipework must use welders and welding procedures qualified to ASME B31.3, and maintain records of the qualifications. Remember that ASME B31.3 qualifications are done in accordance with ASME IX.

Section 8.2.2.1: weld preheating

Preheat is controlled by the requirements of B31.3 and the WPS. It can be used as an alternative to PWHT for P1 and P3 steels (except Mn–Mo steels) if the temperature is maintained at not less than 150 °C.

Preheat can also be used on other steels at the discretion of a piping engineer after consideration of environmental cracking risk (mainly meaning stress corrosion cracking) and toughness (impact strength).

Section 8.2.2.2: postweld heat treatment (PWHT)

This clause contains a list of requirements to be met if you decide to use a local PWHT instead of a full PWHT. In this technique heating bands are wrapped 360° around the workpiece.

Local PWHT may be substituted for 360° banding if:

- A procedure is developed by the piping engineer.
- All relevant factors such as material thickness and properties, NDE, weld joint, welding stresses and distortions, etc., are considered.
- The specified preheat value greater than 150 °C is maintained during welding.
- The required PWHT temperature is maintained at least two material thicknesses (2t) from the weld and monitored by a minimum of two thermocouples.

- Controlled heating is applied to a branch or attachments within the PWHT area.
- The PWHT is for code compliance and not environmental cracking resistance.

Section 8.2.3: design
There are three fairly straightforward requirements here:

- Butt joints have to be *full penetration groove welds*.
- Fillet welded patches must account for weld joint efficiency and crevice corrosion.
- Fillet welded patches must be:
  ○ of equal design strength to a reinforced opening;
  ○ designed to absorb the membrane strain of the part (this means that they must be flexible enough, i.e. not suffer from *restraint* forces).

API 570 appendix C: repairs
One of the most important points here is the welding processes that are allowed to be used for repairs. The ones specifically recommended for repairs are SMAW or GTAW (otherwise known as TIG). These processes were explained previously.

Appendix C contains some important restrictions on SMAW pipework weld repairs (see Fig. 8.1):

- Low-hydrogen electrodes have to be used when the ambient temperature < 10 °C (50 °F) or the material temperature is < 0 °C (32 °F).
- Reinforcing sleeve fillet welds must be welded *upwards* starting from the bottom using electrodes of ⩽ 5/32 in (4 mm) diameter.
- Longitudinal welds on a reinforcing sleeve can use electrodes of ⩽ 3/16 in (4.8 mm) diameter.
- Longitudinal welds on a reinforcing sleeve may require tape or a backing strip to prevent fusion on the pipe wall unless UT confirms that the wall is thick enough to withstand the fusion.

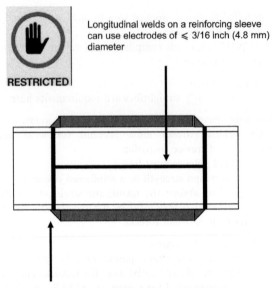

RESTRICTED

Longitudinal welds on a reinforcing sleeve can use electrodes of ⩽ 3/16 inch (4.8 mm) diameter

Reinforcing sleeve fillet welds must be welded upwards, starting from the bottom using smaller electrodes of ⩽ 5/32 inch (4 mm) diameter

Figure 8.1   API 570 SMAW pipe repair restrictions

Other requirements are (see Fig. 8.2):

- Small repair patches should not exceed $\frac{1}{2}$ pipe diameter in any direction.
- Weld small repair patches using electrodes of ⩽ 5/32 in (4 mm) diameter.
- Avoid weaving weld beads when using low-hydrogen electrodes.

Longitudinal welds on a reinforcing sleeve may require tape or a backing strip to prevent fusion on the pipe wall (unless UT confirms the wall is thick enough to withstand the fusion)

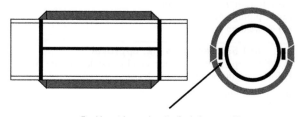

Backing strips on longitudinal sleeve weld

Small repair patches must not exceed ½ pipe diameter in any direction

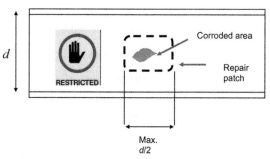

Figure 8.2   More SMAW repair restrictions

8.3   Familiarization questions: API 570 general welding rules

Now try these familiarization questions on the general welding rules of API 570.

Q1. API 570 section 8.1.2: approval

Approval of on-stream welding is required by:

(a) The owner/user ☐
(b) An authorized inspector ☐
(c) A piping engineer ☐
(d) A welding inspector ☐

Q2. API 570 section 8.1.2: approval

Welding repair of in-service cracks should not be attempted before consultation with:

(a) A metallurgist ☐
(b) The piping engineer ☐
(c) An authorized inspector ☐
(d) The design organization ☐

Q3. API 570 section 8.1.3.1: welding repairs: temporary repairs

Temporary repairs should be removed and replaced with a permanent repair at the next available maintenance opportunity or for:

(a) A maximum period of 1 year ☐
(b) A longer period if approved and documented by the inspector ☐
(c) A longer period if approved and documented by the piping engineer ☐
(d) A period not exceeding the MIP relating to the piping class ☐

Q4. API 570 section 8.1.3.1: welding repairs: temporary repairs

A split coupling or patch can be fillet welded on a localized pitted repair area if:

(a) The fillet leg length is equal to the material thickness ☐
(b) The specified minimum yield strength is not less than 275 800 kPa ☐
(c) The specified minimum yield strength is not less than 40 000 psi ☐
(d) The specified minimum yield strength is not more than 40 000 psi ☐

Q5. API 570 section 8.1.3.2: welding repairs: permanent repairs

Flush insert patches can be used to repair damaged or corroded areas:

(a) As long as full penetration groove welds are used ☐
(b) But not on class 1 piping systems ☐
(c) Only on class 2 or class 3 systems ☐
(d) As long as it has rounded corners of less than 1 in (25 mm) radius ☐

Q6. API 570 section 8.2: welding and hot tapping

Any welding conducted on piping components *in operation* must be done in accordance with:

(a) ASME B31.3 ☐
(b) API publication 2201 ☐
(c) ASME IX ☐
(d) API 577 ☐

Q7. API 570 section 8.2.2.1: preheating and PWHT

An alternative to PWHT in certain cases involves preheating:

(a) To not less than 150 °F (70 °C) ☐
(b) All P1 and all P3 steels only ☐
(c) As an alternative to environmental cracking prevention ☐
(d) To not less than 300 °F (150 °C) ☐

Q8. API 570 section 8.2.2.2: preheating and PWHT: PWHT

Local PWHT may be substituted for 360° banding on all materials if:

(a) A minimum preheat of 150 °C or as specified in the WPS is maintained ☐
(b) The PWHT is for code compliance and not environmental crack resistance ☐
(c) Controlled heating is applied to other attachments within the PWHT area ☐
(d) All of the above are relevant ☐

Q9. API 570 section 8.2.3: design

Butt joints welded in accordance with section 8.2.3 can be partial penetration groove welds in the following circumstances

(a) If a repair is temporary ☐

(b) Never ☐
(c) When the inspector permits it ☐
(d) When the completed weld is subjected to 100 % RT or UT ☐

Q10. API 570 section 8.2.3: design

Fillet welded patches require special design considerations and must:

(a) Be designed by the piping engineer ☐
(b) Consider weld joint efficiency ☐
(c) Consider crevice corrosion ☐
(d) Comply with all of the above ☐

8.4 Welding requirements of ASME B31.3 chapter V

B31.3 section 328.5.1: general welding requirements
The main requirements here are:

- All welds shall be done in accordance with a WPS.
- All pressure retaining welds shall be marked with the welder's identification mark.
- Cracked tack welds shall be removed.
- Bridge tacks will be removed.
- Peening is prohibited on the root and final pass.
- Welding must be protected from the weather.

B31.3 section 328.5.2: Fillet and socket welds
Fillet weld shape can vary from convex to concave. Check the following areas in your code:

- Have a look at figure 328.5.2A showing fillet weld sizing.
- Have a look at figure 328.5.2B showing details for slip-on and socket welded flanges.
- Have a look at fig 328.5.2C showing details for components other than flanges.

B31.3 section 328.5.3: seal welds
Seal welds need to cover all exposed threads.

B31.3 section 328.5.4: welded branch connections
This section lays down acceptable requirements for some typical branch connections:

- Figures 328.5.4A to 328.5.4E show details of connections with and without reinforcement pads.

One point worth noting is that if a reinforcing pad or saddle is used then a vent hole must be provided at the side (not the crotch) to:

- reveal leakage between the branch and weld run;
- allow venting during welding and heat treatment.

B31.3 section 328.6: repairs
Repairs have to be carried out by qualified welders using qualified WPSs. In principle, preheating and PWHT need to be the same as for the original weld, unless the PWHT exemption in API 570 is used. **Note:** repairs here relate to initial fabrication repairs and not in-service repairs.

B31.3 section 330: preheating

B31.3 section 330.1: general
This section explains that preheat is used along with PWHT to minimize the detrimental effects of high temperatures and thermal gradients inherent in welding. In fact preheat also retards the cooling rate of the material, and so helps its mechanical properties.

B31.3 section 331: heat treatment
This section explains that PWHT is used to minimize the detrimental effects of high temperatures and thermal gradients inherent in welding and to relieve residual stresses created by bending and forming. Some key points are as follows:

B31.3 section 331.1.1: heat treatment requirements
Table 333.1.1 contains the heat treatment requirements for various material groupings and thicknesses. The WPS will include the heat treatment requirements.

B31.3 section 331.1.3: governing thickness
For welds, the heat treatment is based on the *thicker* of the base materials being joined. Heat treatment is not required for fillet welded flanges and sockets when:

- throat dimension ≤ 16 mm (3/4 in) for P1 material;
- throat dimension ≤ 13 mm (1/2 in) for P3, P4 and P5 with tensile strength ≤ 490 MPa (71 ksi);
- ferritic materials are welded with filler that does not air-harden.

B31.3 section 331.1.7: hardness tests
Hardness tests are carried out to verify satisfactory heat treatment. The hardness limit applies to weld metal and HAZ (close to the weld). Where testing is specified in table 331.1.1 (and this is not for all materials) the requirement is:

- Test a minimum of 10 % of furnace heat-treated components.
- Test 100 % of locally heat-treated components.

B31.3 section 331.2.6: local heat treatment
If heat treatment is localized a circumferential band of the pipe or branch has to be heated to the required temperature. This includes the weldment and an area extending 1 in (25 mm) beyond the ends.

8.5 Familiarization questions: ASME B31.3 general welding rules
Now try these familiarization questions. Find the answers from ASME B31.3 before checking the correct answers at the end of this book.

Q1. ASME B31.3 section 328.1: welding responsibility
When is an employer responsible for welding done by the personnel of his organization?

(a) Always ☐
(b) Except where the WPS is supplied by a third party ☐
(c) Unless it is done in accordance with section 311.2: specific requirements ☐

(d) Unless he subcontracts the final weld acceptance to
another organization ☐

Q2. ASME B31.3 section 328.2.1: welding qualifications

A procedure qualified without a backing ring:

(a) Is not qualified for use with a backing ring in a single
welded butt joint ☐
(b) Is also qualified for use with a backing ring in a single
welded butt joint ☐
(c) Does not require impact testing for procedure
qualification ☐
(d) Is only qualified for use with a backing ring if 100 %
RT is applied ☐

Q3. ASME B31.3 section 328.2.2: procedure qualifications by others

Welding procedures qualified by others may be used:

(a) Subject to specific approval of the piping engineer ☐
(b) As long as the employer is satisfied it meets the
specified requirements ☐
(c) Subject to specific approval of the inspector ☐
(d) But only on base material P1, P3 and P8 ☐

Q4. ASME B31.3 section 328.2.2: performance qualification by others

Can an employer accept a performance qualification made for
another employer?

(a) Yes, but only if the other employer is API certified ☐
(b) Yes, if he obtains a copy of the performance
qualification test certificate ☐
(c) Yes, as long as the inspector is advised within 14 days ☐
(d) Yes, as long as all of the above are satisfied ☐

Q5. ASME B31.3 section 328.3.1: welding materials

Filler metal shall conform to the requirements of ASME IX.
Otherwise approval for its use is required from:

(a) The owner ☐
(b) The inspector ☐
(c) A metallurgist ☐
(d) Any of the above can give approval ☐

Chapter 9

Welding Qualifications and ASME IX

9.1 Introduction
The purpose of this chapter is to familiarize you with the principles and requirements of welding qualification documentation. These are the Weld Procedure Specification (WPS), Procedure Qualification Record (PQR) and Welder Performance Qualification (WPQ). The secondary purpose is to define the essential, non-essential and supplementary essential variables used in qualifying WPSs.

ASME section IX is a part of the ASME boiler pressure vessel code that contains the rules for qualifying welding procedures and welders. It is also used to qualify welders and procedures for welding to ASME B31.3.

9.1.1 Weld procedure documentation: which code to follow?
API 570 (section 8.2.1) states that repair organizations shall use welders and welding procedures qualified to ASME B31.3 and maintain records of the welding procedures and welder performance qualifications. Following this through, ASME B31.3 (section 328.1) states that each employer is responsible for the welding done by the personnel of his organization and shall conduct the tests needed to qualify procedures and welders (but with a few exceptions). It goes on to say (section 328.2) that qualification will conform to the requirements of ASME IX but with a few modifications. These modifications mainly relate to:

- Action that may be taken on failure of an ASME IX 180 degree bend test.
- Preheat and heat treatment requirements to apply in WPS qualification.
- Consumable insert suitability.

- A WPS qualified without backing also qualifies with backing.

ASME IX article II states that each manufacturer and contractor shall prepare written Welding Procedure Specifications (WPS) and a Procedure Qualification Record (PQR), as defined in QW-200.2.

In practice then, API 570 requires the repair organization to carry out welding repairs using welders and procedures that have been qualified in accordance with ASME IX but incorporating the relevant exceptions and modifications specified in ASME B31.3.

9.2 Formulating the qualification requirements

The actions to be taken by the manufacturer to qualify a WPS and welder are done in the following order (see Fig. 9.1).

Step 1: qualify the WPS

- A preliminary WPS (this is an unsigned and unauthorized document) is prepared specifying the ranges of essential variables, supplementary variables (if required) and non-essential variables required for the welding process to be used.
- The required numbers of test coupons are welded and the ranges of essential variables used recorded on the PQR.
- Any required non-destructive testing and/or mechanical testing is carried out and the results recorded in the PQR.

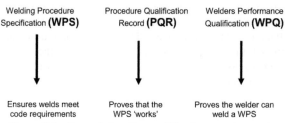

Welding Procedure Specification (WPS)	Procedure Qualification Record (PQR)	Welders Performance Qualification (WPQ)
Ensures welds meet code requirements	Proves that the WPS 'works'	Proves the welder can weld a WPS

Figure 9.1 Formulating the qualification requirements

- If all the above are satisfactory then the WPS is qualified using the documented information on the PQR as proof that the WPS works.

The WPS (see Fig. 9.2) is signed and authorized by the manufacturer for use in production.

Step 2: qualify the welder

The next step is qualify the *welder* by having him weld a test coupon to a qualified WPS. The essential variables used, tests and results are noted and the ranges qualified on a Welder Performance Qualification (WPQ).

Note that ASME IX does not require the use of preheat or PWHT on the welder test coupon. This is because it is the skill of the welder and his ability to follow a procedure that is being tested. The preheat and PWHT are not required because the mechanical properties of the joint have already been determined during qualification of the WPS.

ASME IX QW-250: WPSs and PQRs

We will now look at the ASME IX code rules covering WPS and PQRs (see Fig. 9.3). The code section splits the variables into three groups:

- Essential variables
- Non-essential variables
- Supplementary variables

These are listed on the WPS for each welding process. ASME IX QW-250 lists the variables that must be specified on the WPS and PQR for each process. Note how this is a very long section of the code, consisting mainly of tables covering the different welding processes. There are subtle differences between the approach to each process, but the guiding principles as to what is an essential, non-essential and supplementary variable are much the same.

ASME IX QW 482 WPS format:

Company Name ___MET Ltd___ By: ___S. Hughes___
Welding Procedure Specification No ___SMAW-1___ Date ___01/04/2006___ Supporting PQR No ___SMAW-1___
Revision No ___0___ Date ___01/04/06___

Welding Process(es) ___SMAW___ Type(s) ___Manual___
Automatic, Manual, machine or Semi Automatic

JOINTS (QW-402)

Details

Joint Design ___Single Vee Butt___
Backing (Yes) _____ (No) ___X___ See production drawing
Backing Material (Type) _____
Refer to both backing and retainers

☐ Metal ☐ Nonfusing Metal

☐ Nonmetallic ☐ Other

Sketches, Production Drawings, Weld Symbols or Written Description should show the general arrangement of the parts to be welded. Where applicable, the root spacing and the details of weld groove may be specified. At the option of the manufacturer, sketches may be attached to illustrate joint design, weld layers and bead sequence, eg for notch toughness procedures, for multiple process procedures etc

BASE METALS (QW-403)

P-No. ___1___ Group No. ___2___ to P-No. ___1___ Group No. ___2___
OR
Specification type and grade _____
to
Specification type and grade _____
OR
Chemical Analysis and Mechanical properties _____
to
Chemical Analysis and Mechanical properties _____

Thickness Range:
Base Metal: Groove ___1/16" – 2"___ Fillet _____
Pipe Diameter range: Groove ___All___ Fillet ___All___
Other _____

FILLER METALS (QW-404) Each base metal-filler metal combination should be recorded individually

Spec. No (SFA)	SFA 5.1		
AWS No Class)	E7016		
F No	6		
A-No.	4		
Size of filler metals	All		
Weld Metal			
Thickness range			
Groove	All		
Fillet	All		
Electrode-Flux (Class)	N/A		
Flux Trade Name	N/A		
Consumable Insert	N/A		
Other			

Figure 9.2a   WPS format

ASME IX QW 482 (Back)

WPS No _SMAW-1___ Rev __0___

POSITIONS (QW-405)

Position(s) of Groove _____

Welding Progression: Up _____ Down _____

Position(s) of Fillet _____

PREHEAT (QW-406)

Rate
Preheat Temp. Min _____

Interpass Temp. Max _____

Preheat Maintenance _____
(Continuous or special heating , where applicable, should be recorded)

POSTWELD HEAT TREATMENT (QW-407)

Temperature Range

Time Range

GAS (QW-408)

Percent composition
Gas(es)   (Mixture)   Flow

Shielding _____

Trailing _____

Backing _____

ELECTRICAL CHARACTERISTICS (QW-409)

Current AC or DC _____ Polarity _____
Amps (Range) _____ Volts (Range) _____
(Amps and volts range should be recorded for each electrode size,position, and thickness, etc. this
information may be listed in a tabular form similar to that shown below).

Tungsten Electrode Size and Type_____
(Pure Tungsten, 2% Thoriated, etc)
Mode of Metal Transfer for GMAW _____
(Spray arc, short circuiting arc, etc)
Electrode Wire feed speed range _____

TECHNIQUE (QW-410)

String or Weave Bead _____
Orifice or Gas Cup Size _____
Initial and Interpass Cleaning (Brushing, Grinding, etc) _____

Method of Back Gouging _____
Oscillation _____
Contact Tube to Work Distance _____
Multiple or Single Pass (per side) _____
Multiple or Single Electrodes _____
Travel Speed (Range) _____
Peening _____
Other _____

| Weld Layer(s) | Process | Filler Metal | | Current | | | Travel Speed Range | Other (remarks, comments, hot wire addition, technique, torch angle etc) |
		Class	Diameter	Type Polarity	Amp Range	Volt range		

Figure 9.2b   WPS format

QW 483 PQR format

Company Name _____

Procedure Qualification Record No. _____ Date _____

WPS No _____

Welding Process(es) _____

Types (Manual, Automatic, Semi-Auto)

JOINTS (QW-402)

Groove Design of Test Coupon

(For combination qualifications, the deposited weld metal thickness that shall be recorded for each filler metal or process used)

BASE METALS (QW-403)	**POSTWELD HEAT TREATMENT (QW-407)**
Material spec. _____	Temperature _____
Type or Grade _____	Time _____
P-No _____ to P-No _____	Other _____
Thickness of test coupon _____	
Diameter of test coupon _____	
Other _____	

GAS (QW-408)

Percent composition

Gas(es)   (Mixture)   Flow Rate

Trailing   None _____

Backing   None _____

Backing   None _____

FILLER METALS (QW-404)		**ELECTRICAL CHARACTERISTICS (QW-409)**
SFA Specification _____		
AWS Classification _____		Current _____
Filler metal F-No _____		
Weld Metal Analysis A-No _____		Polarity _____
Size of Filler Metal _____		
Other _____		Amps _____ Volts _____
		Tungsten Electrode Size _____
Weld Metal Thickness _____		Other _____

POSITIONS (QW-405)	**TECHNIQUE (QW-410)**
Position of Groove _____	Travel Speed _____
Welding Progression (Uphill, Downhill) _____	String or Weave Bead _____
Other _____	Oscillation _____
	Multiple or Single Pass (per side) _____
PREHEAT (QW-406)	Single or Multiple Electrodes _____
Preheat Temp. _____	Other _____
Interpass Temp. _____	
Other _____	

Figure 9.3a   PQR format

QW 483 PQR (Back)

PQR No. __SMAW-1

TENSILE TEST (QW-150)

Specimen No	Width	Thickness	Area	Ultimate Total load lb	Ultimate Unit Stress psi	Type of Failure & Location

GUIDED- BEND TESTS (QW-160)

Type and Figure No	Result

TOUGHNESS TESTS (QW-170)

Specimen No	Notch Location	Specimen Size	Test Temp	Impact Values			Drop Weight Break (Y/N)
				Ft-lb	% Shear	Mils	

Comments _____

FILLET WELD TEST (QW-180)

Result – Satisfactory? :   Yes_____ No _____   Penetration into Parent Metal? : Yes_____ No _____

Macro – Results _____

OTHER TESTS

Type of Test _____

Deposit Analysis _____

Other _____

Welder's Name _____Clock No. _____ Stamp No. _____

Tests conducted by: _____Laboratory Test No._____

We certify that the statements in this record are correct and that the tests welds were prepared, welded, and tested in accordance with the requirements of Section IX of the ASME Code.

Manufacturer _____

Date _____   By _____

(Detail of record of tests are illustrative only and may be modified to conform to the type and number of tests required by the Code.)

Figure 9.3b   PQR format

ASME IX QW-350: WPQs

ASME IX QW-350 lists variables by process for qualifying welder performance. These are much shorter and more straightforward than those for WPS/PQRs.

ASME IX welding documentation formats

The main welding documents specified in ASME IX have examples in non-mandatory appendix B section QW-482. Strangely, these are not included in the API 570 exam code document package but fortunately two of them, the WPS and PQR, are repeated in API 577 (have a look at them in API 577 appendix C). Remember that the actual format of the procedure sheets is not mandatory, as long as the necessary information is included.

The other two that are in ASME IX non-mandatory appendix B (the WPQ and Standard Weld Procedure Specification (SWPS)) are not given in API 577 and are therefore peripheral to the API 570 exam syllabus.

9.3   Welding documentation reviews: the exam questions

The main thrust of the API 570 and ASME IX questions is based on the requirement to review a WPS and its qualifying PQR, so these are the documents that you must become familiar with. The review will be subject to the following limitations (to make it simpler for you):

- The WPS and its supporting PQR will contain only *one* welding process.
- The welding process will be SMAW, GTAW, GMAW or SAW and will have only one filler metal.
- The base material P-group number will be either P1, P3, P4, P5 or P8.

Base materials are assigned P-numbers in ASME IX to reduce the amount of procedure qualifications required. The P-number is based on material characteristics like weldability and mechanical properties. S-numbers are the same idea as P-numbers but deal with piping materials from ASME B31.3.

9.3.1 WPS/PQR review questions in the exam

The API 570 certification exam requires candidates to review a WPS and its supporting PQR. The format of these will be based on the sample documents contained in annex B of ASME IX. Remember that this annex B is not contained in your code document package; instead, you have to look at the formats in API 577 appendix B, where they *are* shown (they are exactly the same).

The WPS/PQR documents are designed to cover the parameter/variable requirements of the SMAW, GTAW, GMAW and SAW welding processes. The open-book questions on these documents in the API exam, however, only contain *one* of those welding processes. This means that there will be areas on the WPS and PQR documents that will be left unaddressed, depending on what process is used. For example, if GTAW welding is *not* specified then the details of tungsten electrode size and type will not be required on the WPS/PQR.

In the exam questions, you will need to understand the variables to enable you to determine if they have been correctly addressed in the WPS and PQR for any given process.

9.3.2 Code cross-references

One area of ASME IX that some people find confusing is the numbering and cross-referencing of paragraphs that take place throughout the code. Figure 9.4 explains how the ASME IX numbering system works.

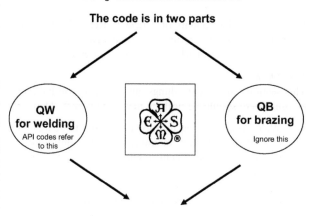

The code is in two parts

QW
for welding
API codes refer
to this

QB
for brazing
Ignore this

Each part is divided into ARTICLES

There are FIVE Welding Articles

Article I   Welding General Requirements
Article II   Welding Procedure Qualifications (WPSs and PQRs)
Article III   Welding Performance Qualifications (WPQs)
Article IV   Welding Data
Article V   Standard Welding Procedure Specifications (SWPS)

Figure 9.4   The ASME IX numbering system

9.4 ASME IX article I

Article I contains less technical 'meat' than some of the following articles (particularly articles II and IV). It is more a collection of general statements than a schedule of firm technical requirements. What it does do, however, is cross-reference a lot of other clauses (particularly in article IV), which is where the more detailed technical requirements are contained.

From the API exam viewpoint, most of the questions that can be asked about article I are:

- more suitable to closed-book questions than open-book ones.
- fairly general and 'common-sense' in nature.

Don't ignore the content of article I. Read the following summaries through carefully but treat article I more as a lead-in to the other articles, rather than an end in itself.

QW-100.1
This section tells you six things, all of which you have met before. There should be nothing new to you here. They are:

- A Weld Procedure Specification (WPS) has to be qualified (by a PQR) by the manufacturer or contractor to determine that a weldment meets its required mechanical properties.
- The WPS specifies the conditions under which welding is performed and these are called welding 'variables'.
- The WPS must address the essential and non-essential variables for each welding process used in production.
- The WPS must address the supplementary essential variables if notch toughness testing is required by other code sections.
- A Procedure Qualification Record (PQR) will document the welding history of the WPS test coupon and record the results of any testing required.

QW-100.2
A welder qualification (i.e. the WPQ) is to determine a welder's ability to deposit sound weld metal or a welding operator's mechanical ability to operate machine-welding equipment.

QW-140: Types and purposes of tests and examinations

QW-141: mechanical tests
Mechanical tests used in procedure or performance qualification are as follows:

- *QW-141.1: tension tests* (see Fig. 9.5). Tension tests are used to determine the strength of groove weld joints.
- *QW-141.2: guided-bend tests* (see Fig. 9.6). Guided-bend tests are used to determine the degree of soundness and ductility of groove-weld joints.

- *QW-141.3: fillet-weld tests.* Fillet weld tests are used to determine the size, contour and degree of soundness of fillet welds.
- *QW-141.4: notch-toughness tests.* Tests are used to determine the notch toughness of the weldment.

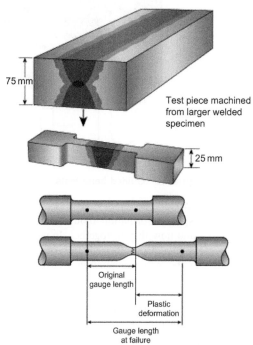

75 mm

Test piece machined from larger welded specimen

25 mm

Original gauge length

Plastic deformation

Gauge length at failure

Figure 9.5   Weld tension tests

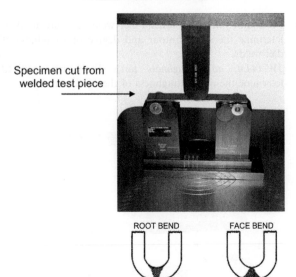

Specimen cut from welded test piece →

ROOT BEND   FACE BEND

ROOT   FACE

Figure 9.6   Guided bend tests

9.5   ASME IX article II

Article II contains hard information about the content of the WPS and PQRs and how they fit together. In common with article I, it cross-references other clauses (particularly in article IV). From the API examination viewpoint there is much more information in here that can form the basis of open-book questions, i.e. about the reviewing of WPS and PQR. ASME IX article II is therefore at the core of the API examination requirements.

QW-200: general
This gives lists of (fairly straightforward) requirements for the WPS and PQR:

- **QW-200.1** covers the WPS. It makes fairly general 'principle' points that you need to understand (but not remember word-for-word).

- **QW-200.2** covers the PQR again. It makes fairly general 'principle' points that you need to understand (but not remember word-for-word).
- **QW-200.3: P-numbers.** These are assigned to base metals to reduce the number of welding procedure qualifications required. For steel and steel alloys, *group* numbers are assigned additionally to P-numbers for the purpose of procedure qualification where notch-toughness requirements are specified.

9.6   Familiarization questions: ASME IX articles I and II

Now try these familiarization questions, using ASME IX articles I and II to find the answers.

Q1.   ASME IX section QW-153: acceptance criteria: tension tests

Which of the following is a true statement on the acceptance criteria of tensile tests?

(a) They must never fail below the UTS of the base material ☐
(b) They must fail in the base material ☐
(c) They must not fail more than 5 % below the minimum UTS of the base material ☐
(d) They must fail in the weld metal otherwise they are discounted ☐

Q2.   ASME IX section QW-200: PQR

A PQR is defined as?

(a) A record supporting a WPS ☐
(b) A record of the welding data used to weld a test coupon ☐
(c) A Procedure Qualification Record ☐
(d) A Provisional Qualification Record ☐

Q3.   ASME IX section QW-200.2(b)

Who certifies the accuracy of a PQR?

(a) The authorized inspector before it can be used ☐
(b) The manufacturer or his designated subcontractor ☐
(c) An independent third party organization ☐
(d) Only the manufacturer or contractor ☐

Q4.   ASME IX section QW-200.3
What is a P-number?

(a) A number assigned to base metals ☐
(b) A procedure unique number ☐
(c) A number used to group similar filler material types ☐
(d) A unique number designed to group ferrous materials ☐

Q5.   ASME IX section QW-200.3
What does the assignment of a group number to a P-number indicate?

(a) The base material is non-ferrous ☐
(b) Postweld heat treatment will be required ☐
(c) The base material is a steel or steel alloy ☐
(d) Notch toughness requirements are mandatory ☐

Q6.   ASME IX section QW-202.2: types of test required
What types of mechanical tests are required to qualify a WPS on full penetration groove welds with no notch toughness requirement?

(a) Tension tests and guided bend tests ☐
(b) Tensile tests and impact tests ☐
(c) Tensile, impact and nick-break tests ☐
(d) Tension, side bend and macro tests ☐

Q7.   ASME IX section QW-251.1
The 'brief of variables' listed in tables QW-252 to QW-265 reference the variables required for each welding process. Where can the complete list of variables be found?

(a) In ASME B31.3 ☐
(b) In ASME IX article IV ☐
(c) In API 570 ☐
(d) In ASME IX article V ☐

Q8.   ASME IX section QW-251.2
What is the purpose of giving base materials a P-number?

(a) It makes identification easier ☐
(b) It reduces the number of welding procedure
    qualifications required ☐
(c) It shows they are in pipe form ☐
(d) It indicates the number of positions it can be welded in ☐

Q9. ASME IX section QW-251.2

A welder performance test is qualified using base material with an S-number. Which of the following statements is true?

(a) Qualification using an S-number qualifies corresponding S- number materials only ☐

(b) Qualification using an S-number qualifies corresponding F-number materials ☐

(c) Qualification using an S-number qualifies corresponding P-number materials only ☐

(d) Qualification using an S-number qualifies *both* P-number and S-number materials ☐

Q10. ASME IX section QW-253

Which of the following would definitely not be a variable consideration for the SMAW process?

(a) Filler materials ☐
(b) Electrical characteristics ☐
(c) Gas ☐
(d) PWHT ☐

9.7 ASME IX article III

Remember that WPQs are specific to the *welder*. Although the content of this article is in the API 570 syllabus it is fair to say that it commands less importance than article II (WPSs and PQRs and their relevant QW-482 and QW-483 format forms) and article IV (welding data).

QW-300.1

This article lists the welding processes separately, with the essential variables that apply to welder and welding operator performance qualifications. The welder qualification is limited by the essential variables listed in QW-350, and defined in article IV: welding data, for each welding process. A welder or welding operator may be qualified by radiography of a test coupon or his initial production welding, or by bend tests taken from a test coupon.

Look at these tables below and mark them with post-it notes:

- Table QW-353 gives SMAW essential variables for welder qualification.
- Table QW-354 gives SAW essential variables for welder qualification.
- Table QW-355 gives GMAW essential variables for welder qualification.
- Table QW-356 gives GTAW essential variables for welder qualification.

QW-351: variable for welders (general)
A welder needs to be requalified whenever a change is made in one or more of the essential variables listed for each welding process.

The limits of deposited weld metal thickness for which a welder will be qualified are dependent upon the thickness of the weld deposited with each welding process, exclusive of any weld reinforcement.

In production welds, welders may not deposit a thickness greater than that for which they are qualified.

9.8   ASME IX article IV
Article IV contains core data about the welding variables themselves. Whereas article II summarizes which variables are essential/non-essential/supplementary for the main welding processes, the content of article IV explains what the variables actually *are*. Note how variables are subdivided into *procedure* and *performance* aspects.

QW-401: general
Each welding variable described in this article is applicable as an essential, supplemental essential or non-essential variable for procedure qualification when referenced in QW-250 for each specific welding process. Note that a change from one welding process to another welding process is an essential variable and requires requalification.

QW-401.1: essential variable (procedure)
This is defined as a change in a welding condition that will affect the mechanical properties (other than notch toughness) of the weldment (e.g. change in P-number, welding process, filler metal, electrode, preheat or postweld heat treatment, etc.).

QW-401.2: essential variable (performance)
A change in a welding condition that will affect the ability of a welder to deposit sound weld metal (such as a change in welding process, electrode F-number, deletion of backing, technique, etc.).

QW-401.3: supplemental essential variable (procedure)
A change in a welding condition that will affect the notch-toughness properties of a weldment (e.g. change in welding process, uphill or down vertical welding, heat input, preheat or PWHT, etc.).

QW-401.4: non-essential variable (procedure)
A change in a welding condition that will not affect the mechanical properties of a weldment (such as joint design, method of back gouging or cleaning, etc.).

QW-401.5
The welding data include the welding variables grouped as follows:

- QW-402: joints
- QW-403: base metals
- QW-404: filler metal
- QW-405: position
- QW-406: preheat
- QW-407: postweld heat treatment
- QW-408: gas
- QW-409: electrical characteristics
- QW-410: technique

QW-420.1: P-numbers
P-numbers are groupings of base materials of similar properties and usability. This grouping of materials allows a reduction in the number of PQRs required. Ferrous P-number metals are assigned a group number if notch toughness is a consideration.

QW-420.2: S-numbers (non-mandatory)
S-numbers are similar to P-numbers but are used for materials not included within ASME BPV Code Material Specifications (section II). There is no mandatory requirement that S-numbers have to be used, but they often are. Note these two key points:

- For WPS a P-number qualifies the same S-number but not vice versa.
- For WPQ a P-number qualifies the same S-number and vice versa.

QW-430: F-numbers
F-number grouping of electrodes and welding rods is based essentially on their usability characteristics. This grouping is made to reduce the number of welding procedure and performance qualifications, where this can logically be done.

QW-432.1
Steel and steel alloys utilize F-number 1 to F-number 6 and are the most commonly used ones.

QW-492: Definitions
QW-492 contains a list of definitions of the common terms relating to welding and brazing that are used in ASME IX.

9.9 Familiarization questions: ASME IX articles III and IV
Try these ASME IX articles III and IV familiarization questions. You will need to refer to your code to find the answers.

Q1. ASME IX section QW-300

What does ASME IX article III contain?

(a) Welding performance qualification requirements ☐
(b) A list of welding processes with essential variables
applying to WPQ ☐
(c) Welder qualification renewals ☐
(d) All of the above ☐

Q2. ASME IX section QW-300.1

What methods can be used to qualify a welder?

(a) By visual and bend tests taken from a test coupon ☐
(b) By visual and radiography of a test coupon or the initial
production weld ☐
(c) By visual, macro and fracture test ☐
(d) Any of the above can be used depending on joint type ☐

Q3. ASME IX section QW-301.3

What must a manufacturer or contractor *not* assign to a qualified
welder to enable his work to be identified?

(a) An identifying number ☐
(b) An identifying letter ☐
(c) An identifying symbol ☐
(d) Any of the above can be assigned ☐

Q4. ASME IX section QW-302.2

If a welder is qualified by radiography, what is the minimum
length of coupon required?

(a) 12 in (300 mm) ☐
(b) 6 in (150 mm) ☐
(c) 3 in (75 mm) ☐
(d) 10 in (250 mm) ☐

Q5. ASME IX section QW-302.4

What areas of a pipe test coupon require visual inspection for a
WPQ?

(a) Inside and outside of the entire circumference ☐
(b) Only outside surface if radiography is to be used ☐
(c) Only the weld metal on the face and root ☐
(d) Visual inspection is not required for pipe coupons ☐

Q6. ASME IX section QW-304
For a WPQ which of the following welding processes can *not* have groove welds qualified by radiography?

(a) GMAW (short circuiting transfer mode) ☐
(b) GTAW ☐
(c) GMAW (globular transfer mode) ☐
(d) They can all be qualified by radiography ☐

Q7. ASME IX section QW-322
How long does a welder's performance qualification last if he has not been involved in production welds using the qualified welding process?

(a) 6 months ☐
(b) 2 years ☐
(c) 3 months ☐
(d) 6 weeks ☐

Q8. ASME IX section QW-402: joints
A welder qualified in a single welded groove weld with backing must requalify if:

(a) He must now weld without backing ☐
(b) The backing material has a nominal change in its composition ☐
(c) There is an increase in the fit-up gap beyond that originally qualified ☐
(d) Any of the above occur ☐

Q9. ASME IX section QW-409.8
What process requires the electrode wire feed speed range to be specified?

(a) SMAW ☐
(b) SAW ☐
(c) GMAW ☐
(d) This term is not used in ASME IX ☐

Q10. ASME IX section QW-416
Which of the following variables would not be included in a WPQ?

(a) Preheat ☐
(b) PWHT ☐
(c) Technique ☐
(d) All of them ☐

9.10   The ASME IX review methodology

One of the major parts of all the API in-service inspection examinations is the topic of weld procedure documentation review. In addition to various 'closed-book' questions about welding processes and techniques, the exams always include a group of 'open-book' questions centred around the activity of checking a Weld Procedure Specification (WPS) and Procedure Qualification Record (PQR).

Note the two governing principles of API examination questions in this subject:

- The PQR and WPS used in exam examples will only contain one welding process and filler material.
- You need only consider essential and non-essential variables (you can ignore supplementary variables).

The basic review methodology is divided into five steps (see Fig. 9.7).

Note the following points to remember as you go through the checklist steps of Fig. 9.7:

- The welding *process* is an *essential* variable and is likely to be SMAW, GTAW, GMAW or SAW.
- Non-essential variables do not have to be recorded in the PQR (but may be at the manufacturer's discretion) and must be addressed in the WPS.
- Information on the PQR will be *actual values* used whereas the WPS may contain a *range* (e.g. base metal *actual* thickness shown in a PQR may be $\frac{1}{2}$ in, while the base metal thickness *range* in the WPS may be 3/16 in to 1 in).
- The process variables listed in tables QW-252 to QW-265 are referred to as the 'brief of variables' and must *not* be used on their own. You *must* refer to the full variable requirements referenced in ASME IX article 4 otherwise you will soon find yourself in trouble.
- The base material will be either P1, P3, P4, P5 or P8 (base materials are assigned P-numbers in ASME IX to reduce the amount of procedure qualifications required).

STEP 1: variables table

- Find the relevant 'brief of variables' table in article 2 of ASME IX for the specified welding process (for example QW-253 for SMAW). This table shows the relevant essential and non-essential variables for the welding process.

STEP 2: PQR check

- Check that the 'editorial' information at the beginning and at the end of the PQR form is filled in.
- Check that all the relevant **essential** variables are addressed on the PQR and highlight any that are not.

STEP 3: WPS check

- Check that the editorial information at the beginning of the WPS form is filled in and agrees with the information on the PQR.
- Check that all the relevant **essential** variables are addressed on the WPS and highlight any that are not.
- Check that all the relevant **non-essential** variables are addressed on the WPS and highlight any that are not.

STEP 4: range of qualification

- Check that the **range of qualification** for each **essential variable** addressed in the PQR is correct and has been correctly stated on the WPS.

STEP 5: number of tensile and bend tests

- Check that the correct type and number of tensile and bend tests have been carried out and recorded on the PQR.
- Check that the tensile/bend test results are correct.

Figure 9.7   The ASME IX WPS/PQR review methodology

9.11 ASME IX WPS/PQR review: worked example

The following WPS/PQR is for an SMAW process and contains typical information that would be included in an exam question.

Figures 9.8 and 9.9 show the WPS and PQR for an SMAW process. Typical questions are given, followed by their answer and explanation.

Step 1: variables table

Q1. (WPS) The base metal thickness range shown on the WPS:

(a) Is correct ☐
(b) Is wrong – it should be 1/16 to $1\frac{1}{2}$ in ☐
(c) **Is wrong – it should be 3/16 to 2 in (QW-451.1)** ☐
(d) Is wrong – it should be 3/8 to 1in ☐

The welding process is SMAW. Therefore the brief of variables used will be those in table QW-253. Look at table QW-253 and check the brief of variables for base metals (QW-403). Note that QW-403.8 specifies that 'change' of thickness T qualified is an essential variable. Therefore the base material thickness must be addressed on the PQR. When we read QW-403.8 in section IV we see that it refers us to QW-451 for the thickness range qualified. Thus:

- The PQR tells us that under base metals (QW-403) the coupon thickness $T = 1$ in.
- QW-451.1 tells us that for a test coupon of thickness $\frac{3}{4} - 1\frac{1}{2}$ in the base material range qualified on the WPS is 3/16 in to $2T$ (therefore $2T = 2$ in).

The correct answer must therefore be (c).

Q2. (WPS) The deposited weld metal thickness:

(a) Is correct ☐
(b) Is wrong – it should be 'unlimited' ☐
(c) Is wrong – it should be 8 in maximum ☐
(d) **Is wrong – it should be 2 in maximum (QW-451.1)** ☐

WPS

Company Name _____MET Ltd_____ By: ___S. Hughes___
Welding Procedure Specification No __SMAW-1__ Date __01/04/2006_ Supporting PQR No _SMAW-1_
Revision No _____0_____ Date __01/04/06__

Welding Process(es) _____SMAW_____ Type(s) _____Manual_____
Automatic, Manual, machine or Semi Automatic

JOINTS (QW-402)

Details

Joint Design _____Single Vee Butt_____
Backing (Yes) _____ (No) ___X___   See production drawing
Backing Material (Type) _____
Refer to both backing and retainers

☐ Metal   ☐ Nonfusing Metal

☐ Nonmetallic   ☐ Other

Sketches, Production Drawings, Weld Symbols or Written Description should show the general arrangement of the parts to be welded. Where applicable, the root spacing and the details of weld groove may be specified. At the option of the manufacturer, sketches may be attached to illustrate joint design, weld layers and bead sequence, eg for notch toughness procedures, for multiple process procedures etc

BASE METALS (QW-403)

P-No. ___1___ Group No. ___2___ to P-No. ___1___ Group No. ___2___
OR
Specification type and grade _____
to
Specification type and grade _____
OR
Chemical Analysis and Mechanical properties _____
to
Chemical Analysis and Mechanical properties _____

Thickness Range:
Base Metal: Groove __1/16" – 2"__ Fillet _____
Pipe Diameter range: Groove __All__ Fillet ___All___
Other _____

FILLER METALS (QW-404) Each base metal-filler metal combination should be recorded individually

Spec. No (SFA)	SFA 5.1		
AWS No Class)	E7016		
F No	6		
A-No	4		
Size of filler metals	All		
Weld Metal			
Thickness range			
Groove	All		
Fillet	All		
Electrode-Flux (Class)	N/A		
Flux Trade Name	N/A		
Consumable Insert	N/A		
Other			

Figure 9.8a   SMAW worked example (WPS)

WPS(Back)

WPS No _SMAW-1___ Rev ___0___

POSITIONS (QW-405)
Position(s) of Groove _____All_____
Welding Progression: Up __Yes__ Down __Yes__
Position(s) of Fillet _____All_____

POSTWELD HEAT TREATMENT (QW407)
Temperature Range ____None____
Time Range ____None____

PREHEAT (QW-406)

Rate
Preheat Temp. Min _____None_____

Interpass Temp. Max _____None_____

Preheat Maintenance _____None_____

(Continuous or special heating , where applicable, should be recorded)

GAS (QW-408)
Percent composition
Gas(es)   (Mixture)   Flow

Shielding __None__ _____

Trailing __None__ _____

Backing ____None____

ELECTRICAL CHARACTERISTICS (QW-409)
Current AC or DC ____DC____   Polarity____Reverse____
Amps (Range) ____110-120____   Volts (Range) __12-20__
(Amps and volts range should be recorded for each electrode size,position, and thickness, etc. this
information may be listed in a tabular form similar to that shown below).

Tungsten Electrode Size and Type _____N/A_____
                                 (Pure Tungsten, 2% Thoriated, etc)
Mode of Metal Transfer for GMAW ____N/A____
                                 (Spray arc, short circuiting arc, etc)
Electrode Wire feed speed range _____N/A_____

TECHNIQUE (QW-410)
String or Weave Bead _____Both_____
Orifice or Gas Cup Size _____N/A_____
Initial and Interpass Cleaning (Brushing, Grinding, etc)____Brushing, grinding____

Method of Back Gouging _____None_____
Oscillation
                        N/A
Contact Tube to Work Distance ____N/A____
Multiple or Single Pass (per side) ____Multiple pass – no pass greater than ¾"____
Multiple or Single Electrodes ____Multiple____
Travel Speed (Range) _____10 IPM_____
Peening
                        Allowed
Other

Weld Layer(s)	Process	Filler Metal		Current				Other (remarks, comments, hot wire addition, technique, torch angle etc)
		Class	Diameter	Type Polarity	Amp Range	Volt range	Travel Speed Range	

Figure 9.8b   SMAW worked example (WPS)

PQR

Company Name_____MET Ltd_____

Procedure Qualification Record No. _____SMAW-1_____ Date ___19/03/06___

WPS No_____SMAW-1_____

Welding Process(es)_____SMAW_____

Types (Manual, Automatic, Semi-Auto)____Manual____

JOINTS (QW-402)

Single Vee Groove, 60 degree included angle. No backing

Groove Design of Test Coupon

(For combination qualifications, the deposited weld metal thickness that shall be recorded for each filler metal or process used)

BASE METALS (QW-403)	POSTWELD HEAT TREATMENT (QW-407)
Material	Temperature __None__
spec.___SA672___	Time _____None_____
Type or Grade____B70____	Other _____
P-No____1____to P-No____1____	
Thickness of test coupon ____1"____	
Diameter of test coupon ____36"____	
Other _____	**GAS (QW-408)**
_____	Percent composition
	Gas(es) (Mixture) Flow Rate
	Trailing ___None___ _____ _____
	Backing ___None___ _____ _____
	Backing ___None___ _____ _____

FILLER METALS (QW-404)	ELECTRICAL CHARACTERISTICS (QW-409)
SFA Specification ____5.1____	
AWS Classification ___E7018___	Current ____Direct____
Filler metal F-No ____4____	
Weld Metal Analysis A-No ___1___	Polarity ____Reverse____
Size of Filler Metal ____1/8"____	
Other _____	Amps _____100_____ Volts ___10___
	Tungsten Electrode Size ____N/A____
Weld Metal Thickness ____1"____	Other _____

POSITIONS (QW-405)	TECHNIQUE (QW-410)
Position of Groove ____3G____	Travel Speed _____25 IPM_____
Welding Progression (Uphill, Downhill) _____	String or Weave Bead ____String____
Other _____	Oscillation ____N/A____
	Multiple or Single Pass (per side) __multiple__
PREHEAT (QW-406)	Single or Multiple Electrodes ____Single____
Preheat Temp. ____200°F____	Other _____
Interpass Temp. ____650°F____	_____
Other _____	_____

Figure 9.9a SMAW worked example (PQR)

PQR (Back)
PQR No. __SMAW-1__

TENSILE TEST (QW-150)

Specimen No	Width	Thickness	Area	Ultimate Total load lb	Ultimate Unit Stress psi	Type of Failure & Location
T-1	0.750	0.985	0.7387	54100	73236	BF/WM
T-2	0.751	0.975	0.6253	40000	63969	BF/WM

GUIDED- BEND TESTS (QW-160)

Type and Figure No	Result
QW-462.2- FACE	Opening 1/16" long –Acceptable
QW-462.2- ROOT	Acceptable

TOUGHNESS TESTS (QW-170)

Specimen No	Notch Location	Specimen Size	Test Temp	Impact Values Ft-lb	% Shear	Mils	Drop Weight Break (Y/N)

Comments _____

FILLET WELD TEST (QW-180)

Result – Satisfactory? : Yes_____ No _____   Penetration into Parent Metal? : Yes_____ No _____

Macro – Results _____

OTHER TESTS
Type of Test _____

Deposit Analysis _____

Other _____

Welder's Name __Richard Easton_____   Clock No. _____   Stamp
No.___RE2____

Tests conducted by: ___LAB Ltd_____   Laboratory Test No. ___LAB01_____

We certify that the statements in this record are correct and that the tests welds were prepared, welded, and tested in accordance with the requirements of Section IX of the ASME Code.

Manufacturer _____MET Ltd_____

Date_____19/03/06_____   By _____S. Hughes_____

Figure 9.9b   SMAW worked example (PQR)

Look at table QW-253 and note how QW-404.30 'change in deposited weld metal thickness t' is an essential variable (and refers to QW-451 for the maximum thickness qualified). Therefore the weld metal thickness must be addressed in the PQR. Thus:

- PQR under QW-404 filler states weld metal thickness $t = 1$ in.
- QW-451 states that if $t \geqslant \frac{3}{4}$ in then the maximum qualified weld metal thickness $= 2T$, where $T =$ base metal thickness.

The correct answer must therefore be (d).

Q3. (WPS) Check of consumable type. The electrode change from E7018 on the PQR to E7016 on the WPS:
(a) **Is acceptable (QW-432)** ☐
(b) Is unacceptable – it can only be an E7018 on the WPS ☐
(c) Is acceptable – provided the electrode is an E7016 A1 ☐
(d) Is unacceptable – the only alternate electrode is an E6010 ☐

Note how QW-404.4 shows that a change in F-number from table QW-432 is an essential variable. This is addressed on the PQR which shows the E7018 electrode as an F- number 4. Table QW-432 and the WPS both show the E7016 electrode is also an F-number 4.

The correct answer must therefore be (a).

Q4. (WPS) Preheat check. The preheat should read:
(a) 60 °F minimum ☐
(b) **100 °F minimum** ☐
(c) 250 °F minimum ☐
(d) 300 °F minimum ☐

QW-406.1 shows that a decrease of preheat > 100 °F (55 °C) is defined as an essential variable. The PQR shows a preheat of 200 °F, which means the minimum shown on the WPS must be 100 °F and not 'none' as shown.

The correct answer must therefore be (b).

Q5. (PQR) Check of tensile test results. The tension tests results are:
(a) Acceptable ☐
(b) Unacceptable – not enough specimens ☐
(c) **Unacceptable – UTS does not meet ASME IX (QW-422-70 ksi)** ☐
(d) Unacceptable – specimen width incorrect ☐

Note how the tensile test part of the PQR directs you to QW-150. On reading this section you will notice that it directs you to the tension test acceptance criteria in QW-153. This says that the minimum procedure qualification tensile values are found in table QW/QB-422.Checking through the figures for material SA-672 Grade B70 shows a minimum specified tensile value of 70 ksi (70 000 psi) but the PQR specimen T-2 shows a UTS value of 63 969 psi.

The correct answer must therefore be (c).

Q6. (PQR) The bend test results are:
(a) Acceptable ☐
(b) Unacceptable – defect size greater than permitted ☐
(c) **Unacceptable – wrong type and number of specimens (QW-450)** ☐
(d) Unacceptable – incorrect figure number – should be QW-463.2 ☐

The PQR directs you to QW-160 for bend tests. For API exam purposes the bend tests will be *transverse* tests. Note these important sections covering bend tests:

- QW-163 gives acceptance criteria for bend tests.
- QW-451 contains PQR thickness limits and test specimen requirements.
- QW-463.2 refers to performance qualifications.

Check of acceptance criteria: From QW-163, the 1/16 in defect is acceptable so answer (b) is incorrect. QW-463.2 refers to *performance qualifications* so answer (d) is incorrect. *Check of test specimen requirements.* QW-451 contains the PQR thickness limits and test specimen requirements.

Consulting this, we can see that for this material thickness (1 in) there is a requirement for four *side bend* tests.

The correct answer is therefore (c).

Q7. (PQR) Certification of PQRs. To be 'code legal' the PQR must be:
(a) **Certified (QW-201)** ☐
(b) Notarized ☐
(c) Authorized ☐
(d) Witnessed ☐

The requirements for certifying PQRs are clearly shown in QW-201. Note how it says *'the manufacturer or contractor shall certify that he has qualified'* ....

The correct answer must therefore be (a).

Q8. (WPS/PQR) Check of variables shown on WPS/ PQR. Essential variable QW-403.9 has been:
(a) Correctly addressed on the WPS ☐
(b) Incorrectly addressed on the WPS ☐
(c) Not addressed on the PQR ☐
(d) **Both (b) and (c)** ☐

Note how QW-253 defines QW-403.9 't-pass' as an essential variable. It must therefore be included on the PQR *and* WPS. Note how in the example it has been addressed on the WPS (under the QW-410 technique) but has not been addressed on the PQR.

The correct answer must therefore be (d).

Q9. (PQR) Variables shown on WPS/PQR. The position of the groove weld is:
(a) Acceptable as shown ☐
(b) Unacceptable – it is an essential variable not addressed ☐
(c) **Unacceptable – position shown is not for pipe (QW-461.4)** ☐
(d) Both (b) and (c) ☐

Remember that weld positions are shown in QW-461. They are not an essential variable, however, so the weld position is not required to be addressed on the PQR. If it is (optionally) shown on the PQR it needs to be checked to make sure it is correct. In this case the position shown refers to the test position of the plate, rather than the pipe.

The correct answer must therefore be (c).

Q10. (PQR/WPS) Variables. The PQR shows 'string' beads but WPS shows 'both' string and weave beads. This is:

(a) Unacceptable – does not meet code requirements ☐
(b) **Acceptable – meets code requirements (non-essential variable – QW200.1c)** ☐
(c) Acceptable – if string beads are only used on the root ☐
(d) Acceptable – if weave beads are only used on the cap ☐

For SMAW, the type of weld bead used is not specified under QW-410 as an essential variable. This means it is a *non-essential* variable and is not required in the PQR (but can be included by choice remember). QW-200.1c permits changes to non-essential variables of a WPS as long as they are recorded. It is therefore acceptable to specify a string bead in the PQR but record it as 'string and weave' in the WPS.

The correct answer must therefore be (b).

SECTION III: NDE AND OTHER TESTING

Chapter 10

General NDE Requirements: API 570, API 577 and ASME B31.3

10.1 Introduction to API 577

API 577 is a recommended practice to give guidance on the welding inspection and metallurgy aspects encountered in the fabrication and repair of refinery and chemical plant equipment and piping. This section is to familiarize you with the non-destructive examination (NDE) methods contained in API 577 section 9.

NDE is used to check for imperfections in a component without destroying it. There are five main methods of NDE covered in API 577:

- Visual testing: VT
- Penetrant testing: PT
- Magnetic testing: MT
- Radiographic testing: RT
- Ultrasonic testing: UT

Each method can utilize several different techniques requiring different skill levels in their application and interpretation of the results obtained. We will look at the basic NDE methods covered in API 577, as these are the ones likely to be referred to in the API examination.

Section 9.1: discontinuities
API 577 works under the assumption that the purpose of an NDE technique is to find *discontinuities*. Note this terminology; a discontinuity is a *finding*, which may (or may not) be serious enough to be classified as a *defect*. This API use of the term discontinuity is not exactly 100 % consistent with its use in other codes, but is near enough.

Several NDE methods can be used to ensure welds do not contain discontinuities that are in excess of their code acceptance criteria. Have a look at the following tables in section 9 of API 577:

- Table 4 lists the various weld joints and common NDE methods used to inspect them.
- Table 5 lists the detection capabilities of the NDE methods for various discontinuities.
- Table 6 lists the types of discontinuity found in carbon steel and stainless steel, the welding process causing it, the NDE method that can find it and some practical solutions for avoiding it.

Section 9.3: visual testing (VT)
This is the most extensively used NDE method and can include direct or indirect examinations:

- **Direct VT** is conducted when the eye can be placed within 6–24 in (150–600 mm) and at an angle not less than 30° to the surface. Mirrors can be used.
- **Indirect (or remote) VT** uses aids such as fibrescopes or borescopes. The equipment used must have a resolution at least equal to direct VT.

Both methods must be illuminated sufficiently to allow resolution of fine detail and a written procedure addressing the illumination requirements should be in place.

10.2 Magnetic particle examination (MT)

Section 9.4.1: general
MT is used to detect surface breaking or slightly subsurface discontinuities in ferromagnetic materials only (material that can be magnetized) (see Fig. 10.1). The material is magnetized and a magnetic field (or flux) is introduced into the component using a permanent magnet, AC or DC electromagnets, or AC or DC prods.

A coating of white contrast paint is applied to the component and a black magnetic ink containing fine iron particles is applied. The particles will be attracted to any areas where magnetic flux leakage is occurring. The flux leakage (or break in the magnetic field) occurs where discontinuities are present, so the particles assume the defect shape.

Advantages of MT

- Can find slightly subsurface defects (unless using AC prods).
- Simple low-cost equipment.
- Low operator skill level.
- Portable.

Disadvantages of MT

- The material must be ferromagnetic (i.e. it cannot test

Magnetic flux passed through the component (using yoke or prod-type magnets)

Black magnetic ink and white contrast paint shows up the defects

Will detect discontinuities such as:

- Cracks
- Laps
- Seams
- Cold shuts
- Laminations

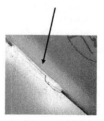

Pie-gauge used to measure magnetic field strength

Figure 10.1   Magnetic testing (MT)

non-magnetic materials such as austenitic stainless steel or aluminium).

- It gives no permanent record.
- Materials may need to be demagnetized after use.
- Using prods can cause arc strikes.

10.3 Liquid penetrant examination (PT) (see Fig. 10.2)

PT can detect discontinuities open to the surface in all materials except porous ones. It is commonly used on non-magnetic materials such as austenitic stainless steel where MT is not possible. A typical colour contrast technique would be carried out as follows:

- The test surface is thoroughly cleaned and degreased.
- The liquid penetrant is applied to the area of concern
- The penetrant is left for a dwell time as recommended by the manufacturer or code (see ASME V requirements later) to give it time to enter any surface-breaking indications by capillary action.
- Excess penetrant is removed from the component surface taking care to prevent the penetrant being washed out of any defects.
- A light coating of the white developer is sprayed on to the component. The penetrant is drawn out of any discontinuities (by a reverse capillary action coupled with a blotting effect from the developer) and stains the developer. An indication of the depth of the discontinuity can be determined from the amount of bleed-out of penetrant from the discontinuity.

Section 9.6.1: liquid penetrant techniques
The two most common techniques are:

- **Colour contrast PT** uses a red penetrant against a white background and requires viewing to take place under good lighting conditions (1000 lux is required under ASME V).
- **Fluorescent PT** uses a dye visible under ultraviolet light

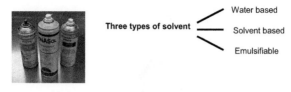

Three types of solvent
- Water based
- Solvent based
- Emulsifiable

Test area

There should be 3 separate aerosols

Figure 10.2 Penetrant testing (PT)

and has to be viewed in a darkened area. This technique is actually the more sensitive and will therefore detect finer linear-type indications than will the colour contrast technique.

10.4 Radiographic inspection (RT)
(see Fig. 10.3)

RT is a *volumetric examination technique* that examines through the entire specimen thickness rather than just on the surface. It is extensively used to check completed welds for surface and subsurface discontinuities. It uses the difference in radiation absorption between solid metal and areas of discontinuity to create an image of differing densities on a radiographic film. Solid metal absorbs more radiation so it hits the film less whereas a void such as a pore will permit more radiation through to reach the film. The special coating on the film causes it to be darker where more radiation hits it.

In the world of API codes an NDE examiner qualified to a minimum of ASNT level II or equivalent performs the interpretation of the results and ensures that the required film

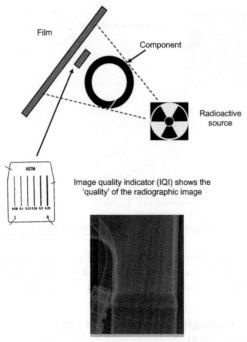

Figure 10.3   Radiographic testing

density and sensitivity has been achieved. The density and sensitivity are designed to ensure that imperfections (i.e. discontinuities) of a dimension relative to the section thickness will be revealed.

Section 9.8.2: image quality indicators
The sensitivity of an industrial radiograph is determined by the use of one or more image quality indicators (IQIs). The IQI will either be a *hole type IQI* containing three different sized holes or a *wire type IQI* containing six different sized wires.

Section 9.8.4: radiographic source selection
The two main methods of radiography are:

- **X-ray radiography** using a large and bulky machine.
- **Gamma radiography** using a radioactive isotope. The common isotopes used are:
  - ○ Iridium 192 for steels between $\frac{1}{4}$ and 3 in (6.3–76.2 mm) thick
  - ○ Cobalt 60 for steels $1\frac{1}{2}$–7 in (38–178 mm) thick

The minimum or maximum thickness that can be radiographed for a given material is determined by demonstrating that the required *sensitivity* has been achieved.

Section 9.8.7: radiographic identification
Identification information has to be plainly and permanently marked on the radiograph but must not obscure any area of interest. Location markers will also appear on the radiograph identifying the area of coverage. This is done using lead markers that appear as white on the radiograph (as the rays cannot penetrate them).

Section 9.8.8: radiographic techniques
One of the most important parts of RT testing is to make sure that the correct technique is used.

The technique is chosen based upon its ability to produce *a good image of suspected discontinuities*, especially the ones not orientated in a favourable direction to the radiographic source. The nature, location and orientation of discontinuities should always be a major factor in establishing the technique. The main techniques are as follows (see Fig. 10.4):

- **The single-wall technique.** A single-wall exposure technique should be used whenever practical. The radiation passes through only one thickness of the material or weld before reaching the film. This gives a *single-wall single-image (SWSI)* radiograph.
- **Double-wall techniques (there are two of these).** If a single-wall technique is impractical then a double-wall technique

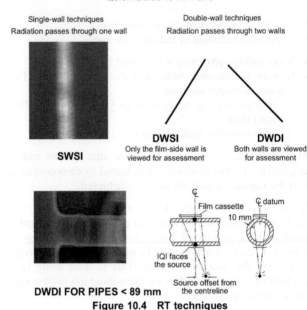

Single-wall techniques
Radiation passes through one wall

Double-wall techniques
Radiation passes through two walls

SWSI

DWSI
Only the film-side wall is
viewed for assessment

DWDI
Both walls are viewed
for assessment

DWDI FOR PIPES < 89 mm

Film cassette
10 mm

IQI faces
the source

Source offset from
the centreline

Figure 10.4   RT techniques

can be used. For welds in component diameters 3.5 in
(88.9 mm) or less a technique can be used (known as the
double-wall double-image (DWDI) technique), which
shows the weld on both walls on the same radiograph.
The source is offset from the weld such that radiation
passes through both walls but gives an elliptical image on
the radiograph for viewing.

A more common technique is the *double-wall single-image
(DWSI) technique*, which positions the source offset on the
wall and means that the radiation passes through both walls
but only leaves one image of a single weld thickness on the
film. A minimum of three exposures taken at 60° or 120°
should be taken for each weld joint.

Section 9.8.9: evaluation of radiographs
The final step in the radiographic process is the evaluation of the radiograph. This should be done by a suitably qualified interpreter who reviews the films and understands the different types of images. The interpreter needs to be aware of:

- The different welding processes and the discontinuities associated with each.
- The difficulty found in detecting planar (non-rounded) discontinuities such as lack of fusion that lie perpendicular to the radiation beam. These give little, if any, change in density on the radiographic image and are therefore difficult to spot.

Section 9.8.9.3: radiographic density
Film *density* is a measure of *how dark the film is* after exposure and processing. It works like this:

- Clear film has a density of zero.
- A density of 1.0 permits 10 % of incident light to pass through it.
- A density of 2.0 permits 1 %.
- A density of 3.0 permits 0.1 %.
- A density of 4.0 permits 0.01 %.

Important point: typical density requirements of the codes are:

- 1.8–4.0 for X-ray and
- 2.0–4.0 for gamma ray.

10.5 Ultrasonic testing (UT)
UT is a specialized inspection technique that requires a high skill level. UT can detect surface and subsurface discontinuities by transmitting a beam of sound in a straight line in the ultrasonic frequency range ($> 20\ 000$ Hz) through the material. If the beam hits a discontinuity it gets reflected back and is amplified to produce an image on a display screen.

Straight beam techniques are used with straight beam (zero degree) compression probes for:

- thickness evaluation or
- lamination checks before angle beams are used to check welds (a lamination could prevent weld discontinuities being found by the angled probes).

The straight beam is directed on to a surface parallel to the contact surface and the time taken for a round trip is displayed. The different display types are:

- The A-scan (see Fig. 10.5)
- The B-scan
- The C-scan

Shear wave (or angle beam) techniques are employed to identify discontinuities in welds. The sound beam enters the weld area at an angle and will either propagate in a straight line or reflect back from a discontinuity, where it will be displayed on a screen. This display enables the operator to determine the size, location and type of discontinuity found.

There are three types of equipment used

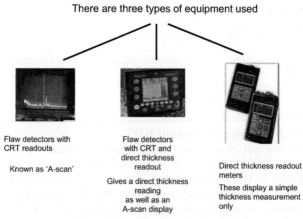

Flaw detectors with
CRT readouts

Known as 'A-scan'

Flaw detectors
with CRT and
direct thickness
readout

Gives a direct thickness
reading
as well as an
A-scan display

Direct thickness readout
meters

These display a simple
thickness measurement
only

Figure 10.5 UT thickness testing

Section 9.9.3: examination coverage
As a general principle of UT, the entire volume of the weld and heat affected zone (HAZ) should be examined. Each pass of the transducer (probe) should overlap the previous pass by 10 % of the transducer dimension and the rate of transducer movement should normally not exceed 6 in (152 mm) per second.

10.6 Hardness testing

Hardness testing of the weld and HAZ is often carried out after PWHT to determine the weldment is in an acceptably 'soft' condition (incorrect PWHT could result in the material having a hardness value that could result in a brittle fracture occurring). Portable hardness testers (such as the Equotip tester) are often used for production welds but can give variable results. Hardness tests performed on test coupons for a PQR in a laboratory will be much more accurate.

10.7 Pressure and leak testing (LT)

Where a hydrostatic or pneumatic pressure test is required, the requirement is simply to follow the relevant code requirements. The testing temperature should be appropriate for the material to avoid brittle fracture. The test must be maintained long enough for a thorough visual inspection to be carried out to identify any potential leaks. Typically, the pressure will be maintained for 30 minutes.

Pneumatic pressure testing often requires special approvals due to the dangers associated with the amount of stored energy in the system. In the UK there is an HSE document called GS4 that gives guidance on pressure testing and pneumatic testing. It gives calculations to determine the thickness of blast barriers required for pneumatic tests. This is not part of the API syllabus.

Leak testing may be carried out to demonstrate system tightness or integrity. ASME section V article 10 addresses leak-testing methods and indicates various test systems to be used for both open and closed units based upon the desired test sensitivity. One of the most common methods is the direct pressure bubble test, where employs a liquid bubble solution applied to areas of a closed system under pressure.

10.8 Familiarization questions: NDE requirements of API 577

Now try these familiarization questions relating to the NDE techniques in API 577.

Q1. API 577 section 9 table 5
Which of the following NDE methods would be unlikely to find an edge breaking lamination in a weld joint?

(a) MT ☐
(b) PT ☐
(c) RT ☐
(d) UT ☐

Q2. API 577 section 9 table 5
Which of the following NDE methods would you need to use if you had welded with a process susceptible to lack of fusion discontinuities?

(a) RT ☐
(b) PT ☐
(c) UT ☐
(d) MT ☐

Q3. API 577 section 9 table 6
Which of the following NDE methods would you need to use if you had welded with the GTAW process and wanted to check for tungsten inclusions?

(a) PT ☐
(b) RT ☐
(c) MT ☐
(d) UT ☐

Q4. API 577 section 9.3.1
What access is required at the surface to carry out a direct visual examination?

(a) The eye must be placed less than 6 in at not less than 30° ☐
(b) The eye must be placed within 6 in at not more than 30° ☐

132

(c) The eye must be placed within 6–24 in at not more than 30° ☐
(d) The eye must be placed within 6–24 in at not less than 30° ☐

Q5. API 577 section 9.9.1.1
In UT, what does DAC stand for?

(a) Distance amplitude correction ☐
(b) Distance at centre ☐
(c) Distance amplitude curve ☐
(d) None of the above ☐

Q6. API 577 section 9.4.1
How would you obtain the best results during an MT examination?

(a) Apply a thick coat of contrast paint to the surface ☐
(b) Carry out two tests perpendicular to each other ☐
(c) Use a dry powder technique ☐
(d) Preheat the test specimen to 52 °C before testing ☐

Q7. API 577 section 9.6.1
Which of the following is not a PT technique?

(a) Colour contrast water washable ☐
(b) Colour contrast solvent removeable ☐
(c) Colour contrast dry powder ☐
(d) Colour contrast post-emulsifiable ☐

Q8. API 577 section 9.8.2
How many holes are found on a hole type IQI?

(a) Six ☐
(b) Four ☐
(c) Five ☐
(d) Three ☐

Q9. API 577 section 9.8.4
What is the API 577 stated thickness range for using iridium 192 in gamma radiography?

(a) 1.0–3.0 in ☐
(b) Above 25 mm ☐
(c) 0.25–5 in ☐
(d) 0.25–3.0 in ☐

Q10. API 577 section 9.8.9.1

A low-power magnification device can be used for viewing radiographs. What would the magnification range be?

(a) 1–2× ☐
(b) Up to 5× ☐
(c) Up to 10× ☐
(d) 1.5–3× ☐

10.9 Introduction to NDE rules of API 570 and ASME B31.3

This section is to familiarize you with the general NDE rules contained in API 570 and ASME B31.3 and compare the requirements of each.

API 570 is for in-service inspections and therefore most NDE carried out will be on repair welds or alterations. The other area requiring NDE will be areas of in-service corrosion or erosion. In reality API 570 says very little about NDE and refers you to the *applicable code* (see API 570 section 8.2.5), which in this case is ASME B31.3.

ASME B31.3 lays down the rules for welding in chapter V, which we covered previously. Chapter VI of B31.3: *Inspection, examination and testing*, however, contains inspection requirements (including weld acceptance criteria) and therefore the rules concerning NDE.

10.10 API 570: NDE rules

API 570 section 8.2.5: non-destructive examination
This states that: *Acceptance of a welded repair or alteration shall include NDE in accordance with the applicable code and the owner/user's specification unless otherwise specified in API 570.* What this really tells us is that it is actually ASME B31.3 that contains the main NDE requirements. API 570 does, however, contain the following specific requirements.

API 570 section 8.1.3.2: permanent repairs
This section basically tells us that insert patches (flush patches) used on Class 1 and Class 2 piping systems must have full-penetration groove welds that are 100 % radio-

graphed or ultrasonically tested using NDE procedures approved by the inspector.

API 570 section 8.2.6: pressure testing
When it is not practical to perform a pressure test of a final closure butt weld joining a replacement piping section to an existing system, the final closure butt weld shall be of 100 % radiographic quality. Notice that the phrase here is 100 % radiographic *quality* and does *not* say it must be 100 % radiographed. Angle-beam UT flaw detection (angle probe/ shear wave techniques) may be used instead of RT, provided the appropriate acceptance criteria have been established and qualified NDE examiners are used. Remember that when API codes mention the NDE *examiner*, they really mean the NDE *technician*.

MT or PT have to be done on the root pass and the completed weld for butt-welds, and on the completed weld only for fillet-welds.

API 570 section 5.10: inspection of welds in-service
This section covers NDE requirements for carrying out piping corrosion surveys (called *profile inspections* in API 570) using radiography to identify wall thinning due to corrosion or erosion. Although finding welding defects is not the purpose of these inspections, they are sometimes found because the original welds were either not subject to RT or UT during construction or were only subject to random or spot radiography. This is especially true on small branch connections that are not normally examined during new construction.

If crack-like imperfections are detected while the piping system is in operation, further inspection with RT and/or UT may be used to assess the magnitude of the imperfection.

As per the general philosophy of API codes, the owner/ user must specify industry-qualified UT examiners when he requires the following:

- detection of interior surface (ID) breaking planar flaws when inspecting from the external surface (OD) or
- where detection, characterization, and/or through-wall sizing is required of planar defects.

Note that planar defects are normally two-dimensional defects such as cracks or lack of fusion. This section uses the terms 'imperfection', 'planar flaw' and 'planar defect' almost interchangeably. Don't let this confuse you; just take them all in this case to be the same thing.

10.11   ASME B31.3: NDE rules

ASME B31.3 chapter VI: inspection, examination and testing

B31.3 gives a wide and fairly comprehensive coverage of the requirements for inspection, examination and testing of welds. The API 570 examination syllabus requires that candidates have an overall knowledge of this chapter. If you review the index of chapter VI it divides clearly into seven main sections:

- Section 340: general requirements for who inspects what: less than half a page
- Section 341: examination: specifies the extent of NDE: approximately 4 pages with figures and tables
- Section 342: personnel requirements: a couple of paragraphs only
- Section 343: documentation procedures: a single paragraph only
- Section 344: the types of examination used: less than 2 pages
- Section 345: leak testing of welds: about 3 pages
- Section 346: records: a couple of paragraphs only

Section 340:   inspection

Section 340.1 to 340.4
Remember that this code distinguishes between *examination* and *inspection*. These sections relate to *inspection* require-

ments. Basically, inspections are carried out for the owner by his inspector and these inspections verify that all the examinations and tests required by the code have been carried out. In effect, this means the inspector represents the owner and can access any place where examinations or testing are carried out to enable him to verify code compliance. There is nothing difficult about that.

Section 341: examination

This is an important section, which is divided into multiple subsections.

Section 341.1: general

Examination applies to quality control functions performed by the manufacturer (for components only), fabricator or erector. An examiner is a person who performs quality control examinations. An example would be an NDE practitioner acting for a fabricator *so do not confuse him with the inspector*.

Section 341.2: responsibility for examination

The manufacturer, fabricator or erector is responsible for performing all required examinations and preparing suitable records of examinations and tests for the inspector's use.

Principle: the API inspector verifies that the suppliers have carried out all required examinations and tests and have prepared records of them.

Section 341.3.2: acceptance criteria

This section contains important figures that could be the subject of open-book examination questions. Figure 341.3.2 shows typical weld imperfections. Acceptance criteria have to at least meet the requirements in para. 344.6.2 for UT of welds and table 341.3.2 which states limits on imperfections for welds.

Section 341.4: extent of required examination

This lays down the extent of examinations required for piping subject to normal fluid service, category D fluid service and severe cyclic conditions. The acceptance criteria are contained in table 341.3.2.

Section 341.4.1: what normal fluid service piping must have
(a) A visual examination consisting of:
- Randomly selected materials and components.
- $\geqslant 5$ % of fabrications (with each welder's work represented).
- 100 % of longitudinal welds; see para. 341.5.1(a) for examination of longitudinal welds required to have a joint factor E_j of 0.90.
- Random examination of mechanical joints (100 % if pneumatic testing is to be performed).

(b) Random RT or UT consisting of:
- $\geqslant 5$ % of circumferential butt and mitre groove welds (with each welder's work represented). In-process VT may be substituted for all or part of the RT or UT on a weld-for-weld basis if authorized by the inspector.
- Circumferential welds with an intersecting longitudinal weld(s) having at least the adjacent 38 mm ($1\frac{1}{2}$ in) of each intersecting weld examined.

(c) Certification provided to the inspector by the examiner showing that all required examinations have been carried out.

Section 341.4.2: what category D fluid service piping must have
A visual examination to meet the acceptance criteria stated in table 341.3.2, for category D fluid service.

Section 341.4.3: what piping under severe cyclic conditions must have
(a) A visual examination carried out as for the normal fluid service but all fabrications, fittings and supports, etc., will be examined to check for features that could lead to fatigue failures.
(b) 100 % RT or UT. Of all circumferential butt and mitre groove welds and all fabricated branch connection welds shown in Figure. 328.5.4E (in-process visual examination

supplemented by appropriate NDE, may be substituted on a weld-for-weld basis if authorized by the inspector). Socket welds and branch connection welds that are not radiographed shall be examined by MT or PT.

(c) Certification provided to the inspector by the examiner showing that all required examinations have been carried out.

Section 341.5.1: spot radiography
American codes frequently use the terminology *spot radiography*. This simply means a sample. This section explains how big the sample has to be in various situations.

(a) For longitudinal groove welds with a weld joint factor (E_j) of 0.90 you have to examine ≥ 300 mm (1 ft) in each 30 m (100 ft) of weld for each welder.

(b) For circumferential groove welds and other welds the requirement is at least one radiograph of 1 in 20 welds for each welder.

Section 344: types of examination
Remember that this is another important section, divided into multiple subsections. All the examination techniques it covers are in the API 570 examination syllabus.

Section 344.2: visual examination
Visual examinations are performed in accordance with ASME V article 9. Records of individual visual examinations are not required, except for those of in-process examination such as welding variables.

Section 344.3: magnetic particle examination
Magnetic particle examination of welds and of components other than castings shall be performed in accordance with ASME V article 7.

Section 344.4: liquid penetrant examination
Liquid penetrant examination of welds and of components shall be performed in accordance with ASME V article 6.

Section 344.5: radiographic examination
Radiography of welds and of components other than castings shall be performed in accordance with ASME V article 2.

Section 344.5.2: extent of radiography
(a) 100 % radiography applies to girth and mitre groove welds and to a fabricated branch connection designed for 100 % radiography, unless otherwise specified in the engineering design.
(b) Random radiography applies to girth and mitre groove welds.
(c) Spot radiography requires a single exposure radiograph at a point within a specified extent of welding. For girth, mitre and branch groove welds the minimum requirement is:
  • Pipes ⩽ DN 65 (NPS 2½), a single elliptical exposure encompassing the entire weld circumference.
  • Pipes > DN 65, the lesser of 25 % of the inside circumference or 152 mm (6 in). For longitudinal welds the minimum requirement is 152 mm (6 in) of weld length.

Section 344.6: ultrasonic examination

Section 344.6.1: method examination
Ultrasonic examination of welds is performed in accordance with ASME V article 5.

10.12 Familiarization questions: API 570 and ASME B31.3 NDE questions (1)

Now try these familiarization questions relating to the NDE requirements of API 570 and ASME B31.3.

Q1. API 570 section 8.2.6: pressure testing

Substituting special procedures for a pressure test after an alteration or repair may be done:

(a) After consultation with the inspector and piping engineer ☐
(b) At the repair organization's discretion ☐
(c) Never. A pressure test is always required ☐
(d) Only with full penetration groove welds ☐

Q2.   API 570 section 8.2.6: pressure testing

Which of the following statements is *not* relevant when it is impractical to perform a pressure test on a final closure weld joining new piping to the existing system?

(a) The new piping will be pressure tested ☐
(b) MT or PT must be performed on the root pass and
face of butt welds ☐
(c) Final closure butt welds must be 100 % radiographic
quality ☐
(d) The final closure weld must *always* be radiographed. ☐

Q3.   API 570 section 8.2.6: pressure testing

Which NDE method will be specified for the detection of internal surface breaking planar flaws in a sealed pipe?

(a) MT ☐
(b) PT ☐
(c) RT ☐
(d) UT ☐

Q4.   API 570 section 8.1.3.2: permanent repairs

A 'class 2' piping system has a flush insert patch welded into it. The inspector scrutinizes the NDE report and determines that it has been tested by the PT method in accordance with API 570 but only on the final weld run. What action should he take?

(a) Call for RT or UT of the final weld ☐
(b) Call for 100 % MT of the final weld run ☐
(c) Final closure butt welds will be 100 % radiographic quality ☐
(d) No action; this is acceptable ☐

Q5.   ASME B31.3 section 341.3.1

What determines the extent of NDE carried out on a piping system before the first use?

(a) The type of butt welds used ☐
(b) The fluid class or loading conditions ☐
(c) The pressure test results ☐
(d) The nominal pipe size (NPS) ☐

Q6. ASME B31.3 table 341.3.2: acceptance criteria

A longitudinal groove weld has been radiographed and contains incomplete penetration of 20 mm in a 150 mm length of weld. What action is required?

(a) None; this is acceptable under ASME B31.3 ☐
(b) Determine the depth of incomplete penetration before
    sentencing it ☐
(c) Reject the weld; incomplete penetration is not permitted ☐
(d) Reject the weld, but *only* for severe cyclic conditions ☐

Q7. ASME B31.3 table 341.3.2: acceptance criteria

A longitudinal groove weld in a 12 mm thick pipe in category D fluid service has been visually examined and has a reinforcement height of 5 mm. What action is required?

(a) None; this is acceptable under ASME B31.3 ☐
(b) Grind the reinforcement fully off and then accept it ☐
(c) Reduce the reinforcement height to a maximum of
    3 mm and then accept it ☐
(d) Reject the weld ☐

Q8. ASME B31.3 section 341.3.1

What is the action to be taken if a spot or random RT examination on a sample reveals a weld defect?

(a) RT two additional samples of the same kind by the
    same welder ☐
(b) UT two additional samples of the same kind by the
    same welder ☐
(c) Immediately remove all similar welds from active service ☐
(d) RT or UT any two further similar samples ☐

Q9. ASME B31.3 section 341.4.1

What would be the *usual* extent of examination for butt welds in *normal class* fluid service piping?

(a) 100 % VT and 100 % RT or UT ☐
(b) Random VT and 100 % RT or UT ☐
(c) Random VT and random RT or UT ☐
(d) 100 % VT and random RT or UT ☐

Q10. ASME B31.3 section 341.4.1

Which of these is true, when applied to normal class fluid service piping?

(a) At least 5 % of all butt welds *must* be subjected to
     RT or UT ☐
(b) In-process VT can be substituted for RT or UT on a
     weld-for-weld basis ☐
(c) At least 38 mm of all longitudinal welds must be
     subjected to RT or UT ☐
(d) The manufacturer will authorize replacement of RT
     or UT with VT ☐

10.13   ASME B31.3 section 345: pressure and leak testing

ASME B31.3 chapter VI section 345 discusses various leak tests that are carried out on piping after weld repairs (and of course during new construction). The main test is the hydrostatic leak test but alternatives are given for when this cannot be carried out for various reasons. The suggested tests are as follows.

10.13.1   The hydrostatic leak test

This is actually incorrectly named as it implies that the pressure is from a static head. Think of it as a *standard hydraulic pressure test*. The test medium is normally water but can be another non-toxic fluid, with a flashpoint above 49 °C (120 °F), if water will freeze or have a detrimental effect on the system.

There is a specific formula for working out test pressure P_T. Test pressure must be:

(a) Not less than $1\frac{1}{2}$ times the design pressure.
(b) For situations where the design temperature is above the test temperature, the minimum test pressure is calculated using the formula:

$$P_T = \frac{1.5\, PS_T}{S}$$

where P_T = minimum test gauge pressure, P = internal design gauge pressure, S_T = stress value at test temperature, S = stress value at design temperature. However, the value of S_T/S should not exceed 6.5. There is also a limitation that P_T must not produce a stress exceeding the material yield stress at the test temperature.

10.13.2 The pneumatic leak test

Pneumatic testing uses a non-toxic, non-flammable gas (usually air) as the test fluid. It therefore has stored energy present and is more dangerous than a hydrostatic test. The test temperature is important and care must be taken to minimize the chance of brittle failure occurring at low temperatures.

The test pressure is 110 % of design pressure. This pressure is not applied all at once; it must be gradually increased to the smaller of 50 % test pressure or 170 kPa (25 psi), at which time a preliminary check is made, including full visual examination of all joints. The pressure is then gradually increased in steps to the test pressure before being reduced to the *design* pressure and held for 10 minutes before examining for leakage.

A pressure relief device set to test pressure plus the lesser of 345 kPa (50 psi) or 10 % of the test pressure has to be fitted as a safety precaution.

10.13.3 The initial service leak test (category D fluid service piping only)

The owner can replace a hydrostatic test with this one, using only the service fluid as the test fluid. During or prior to initial operation, the pressure is gradually increased in steps until the *operating pressure* is reached. A preliminary check is made at half this pressure (or 25 psi if the service fluid is a gas or vapour). Any joints and connections previously tested can be omitted from this test.

10.13.4 The sensitive leak test

This test must be in accordance with the gas and bubble test method specified in ASME V article 10, or by another method demonstrated to have equal sensitivity. Sensitivity of the test has to be not less than 10^{-3} atm ml/s under test conditions.

The test pressure is at least the lesser of 105 kPa (15 psi) gauge or 25 % of the design pressure. The pressure is gradually increased to the lesser of 50 % of the test pressure or 170 kPa (25 psi), at which time a preliminary check is made. Then the pressure is gradually increased in steps until the test pressure is reached. The pressure is held long enough at each step to equalize piping strains.

10.13.5 Alternative leak test

The following alternative method may be used only when the owner considers both hydrostatic and pneumatic leak testing impracticable or too dangerous.

Welds, which have not been subjected to hydrostatic or pneumatic leak tests, shall be examined as follows:

(a) Circumferential, longitudinal and spiral groove welds shall be 100 % RT or UT.
(b) All other welds, including structural attachment welds, shall be examined using PT or MT.
   • A flexibility analysis of the piping system is required beforehand.
   • The system must be subjected to a sensitive leak test.

The purpose of all the aforementioned tests is to ensure *tightness of the piping system*. They are carried out after any NDE requirements but before initial system operation. Note that it is the responsibility of the owner to determine what leak testing needs to be carried out.

10.14 Familiarization questions: ASME B31.3 NDE questions (2)

Now try these familiarization questions on the examination techniques covered by ASME B31.3. Use your code to help you track down the answers.

Q1. ASME B31.3 section 341.4.3

What would be the *usual* extent of examination for butt welds in severe cyclic conditions service piping?

(a) 100 % VT and 100 % RT or UT ☐
(b) Random VT and 100 % RT or UT ☐
(c) Random VT and random RT or UT ☐
(d) 100 % VT and random RT or UT ☐

Q2. ASME B31.3 section 341.5.1

What is the spot radiography requirement for longitudinal groove welds with a joint factor (E_j) of 0.90?

(a) At least 30 mm per 10 m of weld for each welder ☐
(b) At least 300 mm per 100 m of weld for each welder ☐
(c) At least 300 mm per 10 m of weld for each welder ☐
(d) At least 300 mm per 30 m of weld for each welder ☐

Q3. ASME B31.3 section 341.5.1

What is the spot radiography recommendation for circumferential groove welds?

(a) Full RT of more than one weld in 20 for each welder ☐
(b) Not less than one shot in one in 20 welds for each welder ☐
(c) At least 1 foot in every 100 feet of weld for each welder ☐
(d) At least 1 foot in every 10 feet of weld for each welder ☐

Q4. ASME B31.3 section 342.2

Under what circumstances can in-process examinations be carried out by those personnel performing the production work?

(a) When they are SNT-TC-1A qualified ☐
(b) When the manufacturer or installer deems it necessary ☐
(c) When both (a) and (b) above are in place ☐
(d) Under no circumstances ☐

Q5. ASME B31.3 section 345.1

When can an initial service leak test replace the hydrostatic leak test?

(a) When the system is a category D piping system and the
    owner permits it ☐
(b) When the system is a category M piping system and the
    owner permits it ☐
(c) When the system is not under severe cyclic condition ☐
(d) Never; they are complementary tests ☐

Q6. ASME B31.3 section 345.2.1

What is the purpose of a preliminary pneumatic test?

(a) To check for leaks prior to using the system for the first
    time ☐
(b) To check for major leaks using a maximum air pressure
    of 25 psi prior to hydrostatic testing ☐
(c) To check for major leaks using a maximum air pressure
    of 170 psi prior to hydrostatic testing ☐
(d) It is a basic strength test using air as the test fluid ☐

Q7. ASME B31.3 section 345.2.2

What is the minimum time a leak test should be maintained?

(a) 1 hour ☐
(b) 30 minutes ☐
(c) 10 minutes ☐
(d) 5 minutes ☐

Q8. ASME B31.3 section 345.2.3

Under what circumstances can a closure weld be exempted from leak testing?

(a) When the assembly has been leak tested and the weld
    is 100 % VT and RT ☐
(b) When the fabricator uses a coded welder and
    in-service VT ☐
(c) When the system is not category D or M fluid service ☐
(d) Under no circumstances can it be exempted ☐

Q9. ASME B31.3 section 345.4
What is the test fluid required for a hydrostatic leak test?

(a) Any suitable non-toxic liquid ☐
(b) Water ☐
(c) A flammable liquid with a flashpoint of at least 49 °C ☐
(d) Any of the above could be suitable ☐

Q10. ASME B31.3 section 345.5
What is the test pressure for the sensitive leak test?

(a) At least the lower of 105 psi or 25 % of design pressure ☐
(b) At least the lower of 25 kPa or 15 % of design pressure ☐
(c) At least the lower of 15 psi or 25 % of design pressure ☐
(d) At least the lower of 25 psi or 25 % of design pressure ☐

Chapter 11

The NDE Requirements of ASME V

11.1   Introduction

This chapter is to familiarize you with the specific NDE requirements contained in ASME V. ASME B31.3 references ASME V as the supporting code but only articles 1, 2, 6, 7, 9, 10 and 23 are required for use in the API 570 examination. These are the ones on which we will concentrate our efforts.

These articles of ASME V provide the main detail of the NDE techniques that are referred to in many of the API codes. Note that it is only the *body* of the articles that are included in the API examinations; the additional (mandatory and non-mandatory) appendices that some of the articles have are not examinable. We will now look at each of the articles 1, 2, 6, 7, 9, 10 and 23 in turn.

11.2   ASME V article 1: general requirements

Article 1 does little more than set the general scene for the other articles that follow. It covers the general requirement for documentation procedures, equipment calibration and records, etc., but doesn't go into technique-specific detail. Note how the subsections are annotated with T-numbers (as opposed to I-numbers used for the appendices).

Manufacturer versus repairer
One thing that you may find confusing in these articles is the continued reference to *The Manufacturer*. Remember that ASME V is really a code intended for new manufacture. We are using it in its API 570 context, i.e. when it is used to cover repairs. In this context, you can think of The Manufacturer as *The Repairer*.

Table A-110: imperfections and types of NDE method
This table lists imperfections in materials, components and welds and the suggested NDE methods capable of detecting them. Note how it uses the terminology *imperfection*. Some of the other codes would refer to these as discontinuities or indications (yes, it is confusing).

Note how table A-110 is divided into three types of imperfection:

- Service-induced imperfections
- Welding imperfections
- Product form

We are mostly concerned with the service-induced imperfections and welding imperfections because our NDE techniques are to be used with API 570, which deals with in-service inspections and welding repairs.

The NDE methods in table A-110 are divided into those that are capable of finding imperfections that are:

- open to the surface only;
- open to the surface or slightly subsurface;
- located anywhere through the thickness examined.

Note how article 1 provides very basic background information only. The main requirements appear in the other articles, so API examination questions on the actual content of article 1 are generally fairly rare. If they do appear they will probably be closed book, with a very general theme.

11.3 ASME V article 2: radiographic examination

ASME V article 2 covers some of the specifics of radiographic testing techniques. Note that it does not cover anything to do with the *extent* of RT on pipework, i.e. how many radiographs to take or where to do them (we have seen previously that these are covered in ASME B31.3).

Most of article 2 is actually taken up by details of image quality indicators (IQIs) or penetrameters, and parameters such as radiographic density, geometric unsharpness and similar detailed matters. While this is all fairly specialized, it is fair to say that the subject matter lends itself more to open-book exam questions rather than closed-book 'memory' types of questions.

T-210: scope

This explains that article 2 is used in conjunction with the general requirements of article 1 for the examination of materials including castings and welds.

Note that there are seven mandatory appendices detailing the requirements for other product-specific, technique-specific and application-specific procedures. Apart from appendix V, which is a glossary of terms, do not spend time studying these appendices. Just look at the titles and be aware they exist. The same applies to the three non-mandatory appendices.

T-224: radiograph identification

Radiographs have to contain unique traceable permanent identification, along with the identity of the manufacturer and date of the radiograph. The information need not be an image that actually appears on the radiograph itself (i.e. it could be from an indelible marker pen) but usually is.

T-276: IQI (image quality indicator) selection

T-276.1: Material

IQIs have to be selected from either the same alloy material group or an alloy material group or grade with less radiation absorption than the material being radiographed.

Remember that the IQI gives an indication of how 'sensitive' a radiograph is. The idea is that the smallest wire visible will equate to the smallest imperfection size that will be visible on the radiograph.

T-276.2: size of IQI to be used (see Fig. 11.1)

Table T-276 specifies IQI selection for various material thickness ranges. It gives the designated hole size (for hole

Quick Guide to API 570

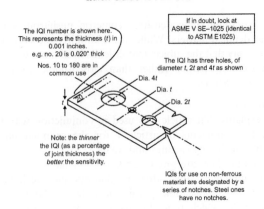

The IQI number is shown here. This represents the thickness (t) in 0.001 inches.
e.g. no. 20 is 0.020" thick

Nos. 10 to 180 are in common use

If in doubt, look at ASME V SE–1025 (identical to ASTM E1025)

The IQI has three holes, of diameter t, $2t$ and $4t$ as shown

Dia. $4t$
Dia. t
Dia. $2t$

Note: the thinner the IQI (as a percentage of joint thickness) the better the sensitivity.

IQIs for use on non-ferrous material are designated by a series of notches. Steel ones have no notches.

Image quality designation is expressed as $(X)-(Y)t$:
(X) is the IQI thickness (t) expressed as a percentage of the joint thickness
$(Y)(t)$ is the hole that must be visible

IQI designation	Sensitivity	Visible hole*
1–2t	1	2t
2–1t	1.4	1t
2–2t	2.0	2t
2–4t	2.8	4t
4–2t	4.0	2t

* The hole that must be visible in order to ensure the sensitivity level shown

T-277.1		ARTICLE 2 — RADIOGRAPHIC EXAMINATION			T-277.2

TABLE T-276
IQI SELECTION

Nominal Single-Wall Material Thickness Range		Source Side		Film Side	
in.	mm	Hole-Type Designation	Wire-Type Essential Wire	Hole-Type Designation	Wire-Type Essential Wire
Up to 0.25, incl.	Up to 6.4, incl.	12	5	10	4
Over 0.25 through 0.375	Over 6.4 through 9.5	15	6	12	5
Over 0.375 through 0.50	Over 9.5 through 12.7	17	7	15	6
Over 0.50 through 0.75	Over 12.7 through 19.0	20	8	17	7
Over 0.75 through 1.00	Over 19.0 through 25.4	25	9	20	8
Over 1.00 through 1.50	Over 25.4 through 38.1	30	10	25	9
Over 1.50 through 2.00	Over 38.1 through 50.8	35	11	30	10
Over 2.00 through 2.50	Over 50.8 through 63.5	40	12	35	11
Over 2.50 through 4.00	Over 63.5 through 101.6	50	13	40	12
Over 4.00 through 6.00	Over 101.6 through 152.4	60	14	50	13
Over 6.00 through 8.00	Over 152.4 through 203.2	80	16	60	14
Over 8.00 through 10.00	Over 203.2 through 254.0	100	17	80	16
Over 10.00 through 12.00	Over 254.0 through 304.8	120	18	100	17
Over 12.00 through 16.00	Over 304.8 through 406.4	160	20	120	18
Over 16.00 through 20.00	Over 406.4 through 508.0	200	21	160	20

Figure 11.1   IQI selection

type IQIs) and the essential wire (for wire type IQIs) when the IQI is placed on either the source side or film side of the weld. Note that the situation differs slightly depending on whether the weld has reinforcement (i.e. a weld cap) or not.

T-277: use of IQIs to monitor radiographic examination

T-277.1: placement of IQIs

For the best results, IQIs are placed on the *source side* (i.e. nearest the radiographic source) of the part being examined. If inaccessibility prevents hand placing the IQI on the source side, *it can be placed on the film side* in contact with the part being examined. If this is done, a lead letter *F* must be placed adjacent to or on the IQI to show it is on the film side. This will show up on the film.

IQI location for welds. Hole type IQIs can be placed adjacent to or on the weld. Wire IQIs are placed on the weld so that the length of the wires is perpendicular to the length of the weld. The identification number(s) and, when used, the lead letter *F* must not be in the area of interest, except where the geometric configuration of the component makes it impractical.

T-277.2: number of IQIs to be used

At least one IQI image must appear on *each radiograph* (except in some special cases). If the radiographic density requirements are met by using more than one IQI, one must be placed in the lightest area and the other in the darkest area of interest. The idea of this is that the intervening areas are then considered as having acceptable density (a sort of interpolation).

T-280: evaluation of radiographs (Fig. 11.2)

This section gives some quite detailed 'quality' requirements designed to make sure that the radiographs are readable and interpreted correctly.

This introduces four parameters that
define the 'quality' of a radiograph

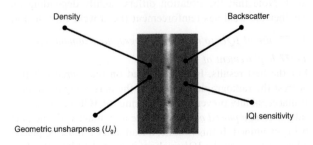

Be prepared to learn about these parameters, and what their values/limits are

Figure 11.2   ASME V article 2 evaluation of radiographs

T-282: radiographic density
These are specific requirements that are based on very well
established requirements used throughout the NDE industry.
It gives numerical values of *density* (a specific measured
parameter), which have to be met for a film to be considered
acceptable.

T-282.1: density limitations
This specifies acceptable density limits as follows:

- Single film with X-ray source: density = 1.8 to 4.0
- Single film with gamma-ray source: density = 2.0 to 4.0
- Multiple films: density = 0.3 to 4.0

A tolerance of 0.05 in density is allowed for variations
between densitometer readings.

T-283: IQI sensitivity

T-283.1: required sensitivity
In order for a radiograph to be deemed 'sensitive enough' to
show the defects of a required size, the following things must
be visible when viewing the film:

This is undesirable stray radiation (because it fogs the image)

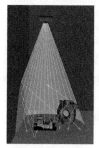

- A lead letter **B** is attached to the back of the film

- If it appears on the image as a light image on a dark background, the image is unacceptable

Figure 11.3   Backscatter gives an unclear image

- For a hole type IQI: the designated hole IQI image and the 2T hole
- For a wire type IQI: the designated wire
- IQI identifying numbers and letters

T-284: excessive backscatter

Backscatter is a term given to the effect of scattering of the X- or gamma rays, leading to an unclear image. If a light image of the lead symbol 'B' appears on a darker background on the radiograph, protection from backscatter is insufficient and the radiograph is unacceptable. A dark image of the 'B' on a lighter background is acceptable (see Fig. 11.3).

T-285: geometric unsharpness limitations

Geometric unsharpness is a numerical value related to the 'fuzziness' of a radiographic image, i.e. an indistinct 'penumbra' area around the outside of the image. It is represented by a parameter U_g (unsharpness due to geometry) calculated from the specimen-to-film distance, focal spot size, etc.

Article 2 section T-285 specifies that geometric unsharpness *(U_g)* of a radiograph shall not exceed the following:

Material thickness, in (mm)	U_g maximum, in (mm)
Under 2 (50.8)	0.020 (0.51)
2 through 3 (50.8–76.2)	0.030 (0.76)
Over 3 through 4 (76.2–101.6)	0.040 (1.02)
Greater than 4 (101.6)	0.070 (1.78)

In all cases, material thickness is defined as the thickness on which the IQI is chosen.

11.4 ASME V article 6: penetrant testing

T-620: general
This article of ASME V explains the principle of penetrant testing (PT). We have already covered much of this in API 577, but ASME V article 6 adds some more formal detail.

T-642: surface preparation before doing PT
Surfaces can be in the as-welded, as-rolled, as-cast or as-forged condition and may be prepared by grinding, machining or other methods as necessary to prevent surface irregularities masking indications. The area of interest, and adjacent surfaces within 1 in (25 mm), need to be prepared and degreased so that indications open to the surface are not obscured.

T-651: the PT techniques themselves
Article 6 recognizes *three penetrant processes*:

- Water washable
- Post-emulsifying (not water based but will wash off with water)
- Solvent removable

The three processes are used in combination with the *two penetrant types* (visible or fluorescent), resulting in a total of six liquid penetrant techniques.

T-652: PT techniques for standard temperatures
For a standard PT technique, the temperature of the penetrant and the surface of the part to be processed must

be between 50 °F (10 °C) and 125 °F (52 °C) throughout the examination period. Local heating or cooling is permitted to maintain this temperature range.

T-670: the PT examination technique (see Fig. 11.4)

T-671: penetrant application

Penetrant may be applied by any suitable means, such as dipping, brushing or spraying. If the penetrant is applied by spraying using a compressed-air type of apparatus, filters have to be placed on the upstream side near the air inlet to stop contamination of the penetrant by oil, water, dirt or sediment that may have collected in the lines.

T-672: penetration time

Penetration time is critical. The minimum penetration time must be as required in table T-672 or as qualified by demonstration for specific applications.

Note: while it is always a good idea to follow the manufacturers' instructions regarding use and dwell times for their penetrant materials, this table T-672 lays down *minimum* dwell times for the penetrant and developer. These are the minimum values that would form the basis of any exam questions based on ASME V.

T-676: interpretation of PT results

T-676.1: final interpretation

Final interpretation of the PT results has to be made within 10 to 60 minutes after the developer has dried. If bleed-out does not alter the examination results, longer periods are permitted. If the surface to be examined is too large to complete the examination within the prescribed or established time, the examination should be performed in increments.

This is simply saying: *inspect within 10–60 minutes*. A longer time can be used if you expect very fine imperfections. Very large surfaces can be split into sections.

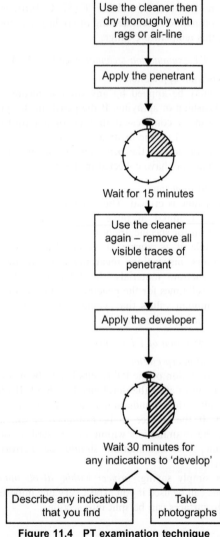

Figure 11.4 PT examination technique

T-676.2: characterizing indication(s)

Deciding (called *characterizing* in ASME-speak) the types of discontinuities can be difficult if the penetrant diffuses excessively into the developer. If this condition occurs, close observation of the formation of indications during application of the developer may assist in characterizing and determining the extent of the indications. In other words, the shape of deep indications can be masked by heavy leaching out of the penetrant, so it is advisable to start the examination of the part as soon as the developer is applied.

T-676.4: fluorescent penetrants

With fluorescent penetrants, the process is essentially the same as for colour contrast, but the examination is performed using an ultraviolet light, sometimes called *black light*. This is performed as follows:

(a) It is performed in a darkened area.
(b) The examiner must be in the darkened area for at least 5 minutes prior to performing the examination to enable his eyes to adapt to dark viewing. He must not wear photosensitive glasses or lenses.
(c) Warm up the black light for a minimum of 5 min prior to use and measure the intensity of the ultraviolet light emitted. Check that the filters and reflectors are clean and undamaged.
(d) Measure the black light intensity with a black light meter. A minimum of 1000 $\mu W/cm^2$ on the surface of the part being examined is required. The black light intensity must be re-verified at least once every 8 hours, whenever the workstation is changed or whenever the bulb is changed.

T-680: evaluation of PT indications

Indications are evaluated using the relevant code acceptance criteria (e.g. B31.3 for pipework). Remember that ASME V does not give acceptance criteria. Be aware that false indications may be caused by localized surface irregularities.

Broad areas of fluorescence or pigmentation can mask defects and must be cleaned and re-examined.

11.5 Familiarization questions: ASME V articles 1, 2 and 6

Now try these familiarization questions on ASME V articles 1, 2 and 6.

Q1. ASME section V article 2: radiography T-223

When performing a radiograph, where is the 'backscatter indicator' lead letter B placed?

(a) On the front of the film holder ☐
(b) On the outside surface of the pipe ☐
(c) On the internal surface of the pipe ☐
(d) On the back of the film holder ☐

Q2. ASME section V article 2: radiography T277.1 (d)

Wire IQIs must be placed so that they are:

(a) At 45° to the weld length ☐
(b) Parallel to the weld metal's length ☐
(c) Perpendicular to the weld metal's longitudinal axis but not across the weld ☐
(d) Perpendicular to the weld metal's longitudinal axis and across the weld ☐

Q3. ASME section V article 6: penetrant testing T-620

Liquid penetrant testing can be used to detect:

(a) Subsurface laminations ☐
(b) Internal flaws ☐
(c) Surface and slightly subsurface discontinuities ☐
(d) Surface breaking discontinuities ☐

Q4. ASME section V article 1: T-150

When an examination to the requirements of section V is required by a code such as ASME B31.3 the responsibility for establishing NDE procedures lies with:

(a) The inspector ☐
(b) The examiner ☐
(c) The user's quality department ☐
(d) The installer, fabricator or manufacturer/repairer ☐

Q5.   ASME section V article 6: penetrant testing mandatory appendix II

Penetrant materials must be checked for the following contaminants when used on austenitic stainless steels:

(a) Chlorine and sulfur content ☐
(b) Fluorine and sulfur content ☐
(c) Fluorine and chlorine content ☐
(d) Fluorine, chlorine and sulfur content ☐

11.6   ASME V article 7: magnetic testing (MT)

Similar to the previous article 6 covering penetrant testing, this article 7 of ASME V explains the technical principle of magnetic testing (MT). As with PT, we have already covered much of this in API 577, but article 7 adds more formal detail. Remember again that it is not component-specific; it deals with the MT techniques themselves, not the *extent* of MT you have to do on a pipework system.

T-720: general

MT methods are used to detect cracks and other discontinuities on or near the surfaces of ferromagnetic materials. It involves magnetizing an area to be examined and then applying ferromagnetic particles to the surface, where they form patterns where the cracks and other discontinuities cause distortions in the normal magnetic field.

Maximum sensitivity is achieved when linear discontinuities are orientated *perpendicular to the lines of magnetic flux*. For optimum effectiveness in detecting all types of discontinuities, each area should therefore be examined at least *twice*, with the lines of flux during one examination approximately perpendicular to the lines of flux during the other, i.e. you need two field directions to do the test properly.

T-750: The MT techniques (see Fig. 11.5)

One or more of the following five magnetization techniques can be used:

(a) Prod technique
(b) Longitudinal magnetization technique

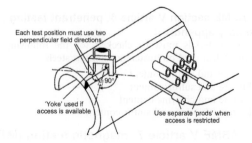

Each test position must use two perpendicular field directions

90°

'Yoke' used if access is available

Use separate 'prods' when access is restricted

MT prod and yoke methods

Figure 11.5   MT examination techniques

(c) Circular magnetization technique
(d) Yoke technique
(e) Multidirectional magnetization technique

The API examination will be based on the prod or yoke techniques (i.e. (a) or (d) above) so these are the only ones we will consider. The others can be ignored for exam purposes.

T-752: the MT prod technique

T-752.1: the magnetizing procedure
Magnetization is accomplished by pressing portable prod type electrical contacts against the surface in the area to be examined. To avoid arcing, a remote control switch, which may be built into the prod handles, must be provided to allow the current to be turned on *after* the prods have been properly positioned.

T-752.3: prod spacing
Prod spacing must not exceed 8 in (203 mm). Shorter spacing may be used to accommodate the geometric limitations of the area being examined or to increase the sensitivity, but prod spacings of less than 3 in (76 mm) are usually not practical due to 'banding' of the magnetic particles around the prods.

The prod tips must be kept clean and dressed (to give good contact).

T-755: the MT yoke technique

This method must only be used (either with AC or DC electromagnetic yokes or permanent magnet yokes) to detect discontinuities that are *surface breaking* on the component.

T-764.1: magnetic field strength

When doing an MT test, the applied magnetic field must have sufficient strength to produce satisfactory indications, but it must not be so strong that it causes the masking of relevant indications by non-relevant accumulations of magnetic particles. Factors that influence the required field strength include:

- Size, shape and material permeability of the part
- The magnetization technique
- Coatings
- The method of particle application
- The type and location of discontinuities to be detected

Magnetic field strength can be verified by using one or more of the following three methods:

- Method 1: T-764.1.1: pie-shaped magnetic particle field indicator
- Method 2: T-764.1.2: artificial flaw shims
- Method 3: T-764.1.3: Hall effect tangential-field probe

T-773: methods of MT examination (dry and wet)

Remember the different types of MT technique. The ferromagnetic particles used as an examination medium can be either *wet* or *dry*, and may be either fluorescent or colour contrast:

- For dry particles, the magnetizing current *remains on* while the examination medium is being applied and excess of the examination medium is removed. Remove the excess particles with a light air stream from a bulb, syringe or air hose (see T-776).

- For wet particles the magnetizing current will be *turned on after applying the particles*. Wet particles fraom aerosol spray cans may be applied before and/or after magnetization. Wet particles can be applied during magnetization as long as they are not applied with sufficient velocity to dislodge accumulated particles.

T-780: Evaluation of defects found during MT testing
As with the other NDE techniques described in ASME V, defects and indications are evaluated using the relevant code acceptance criteria (e.g. ASME B31.3). Be aware that false indications may be caused by localized surface irregularities. Broad areas of particle accumulation can mask relevant indications and must be cleaned and re-examined.

11.7   ASME V article 9: visual examination

This is a fairly short article setting out various requirements to be followed when doing visual examinations of a component. Visual examination is typically carried out when the reference code (e.g. ASME B31.3) does not require any other form of NDE. Most of this article contains common sense rather than any special technical requirements.

T-950: visual examination techniques (Fig. 11.6)
Visual examination is generally used to determine such things as the surface condition of the part, alignment of mating surfaces, shape or evidence of leaking.

T-952: direct visual examination
A summary of requirements for direct (or *close*) visual examination are:

- Your eye must be within 24 in (610 mm) of the surface to be examined and at an angle not less than 30° to the surface.
- Mirrors may be used to improve the angle of vision, and

Close (or direct) visual examination

Remote visual examination

At a distance, i.e. using:

- Borescope
- Mirrors
- Remote cameras

Figure 11.6 ASME V article 9: visual examination techniques

aids such as a magnifying lens may be used to assist examinations.

- Illumination to a minimum light level of 100 foot-candles (1000 lux) is required.

T-953: remote visual examination

In some cases, remote visual examination may have to be substituted for direct (close) examination. Remote visual examination may use visual aids such as mirrors, telescopes, borescopes, fibre optics, cameras or other suitable instruments. Such systems need to have a resolution capability at least equivalent to that obtainable by direct visual observation.

T-980: evaluation of visual results

T-980.1 acceptance code

As for all the other NDE techniques, visual acceptance criteria are set out in the reference code ASME B31.3, rather than here in ASME V.

11.8 Familiarization questions: ASME V articles 7 and 9

Now try these familiarization questions covering ASME V articles 7 and 9.

Q1. ASME section V article 7: magnetic particle testing T-720

Magnetic particle testing can be used to find:

(a) Surface and near-surface discontinuities in all materials ☐
(b) Surface and near-surface discontinuities in ferromagnetic materials ☐
(c) Surface and near-surface discontinuities in all metallic materials ☐
(d) Surface breaking discontinuities only ☐

Q2. ASME section V article 7: magnetic particle testing T-720

During an MT procedure, maximum sensitivity for finding discontinuities will be achieved if:

(a) The lines of magnetic flux are perpendicular to a linear discontinuity ☐
(b) The lines of magnetic flux are perpendicular to a volumetric discontinuity ☐
(c) The lines of magnetic flux are parallel to a linear discontinuity ☐
(d) The lines of magnetic flux are parallel to a volumetric discontinuity ☐

Q3. ASME section V article 7: magnetic particle testing T-741.1(b)

Surfaces must be cleaned of all extraneous matter prior to magnetic testing. How far back must adjacent surfaces to the area of interest be cleaned?

(a) At least 2 inches ☐
(b) At least $\frac{1}{2}$ inch ☐
(c) Cleaning is not required on adjacent surfaces ☐
(d) At least 1 inch ☐

Q4. ASME section V article 7: magnetic particle testing T-741.1(d)

According to ASME V, what is the maximum coating thickness permitted on an area to be examined by MT?

(a) 50 μm ☐
(b) No coating is permitted ☐
(c) 40 μm ☐
(d) An actual value is not specified ☐

Q5. ASME section V article 7: magnetic particle testing T-764.1

Which of the following methods can verify the adequacy of magnetic field strength?

(a) A pie-shaped magnetic particle field indicator ☐
(b) Artificial flaw shims ☐
(c) A gaussmeter and Hall effect tangential-field probe ☐
(d) They can all be used ☐

Q6. ASME section V article 7: magnetic particle testing T-762(c)

What is the lifting power required of a DC electromagnet or permanent magnet yoke?

(a) 40 lb at the maximum pole spacing that will be used ☐
(b) 40 lb at the minimum pole spacing that will be used ☐
(c) 18.1 lb at the maximum pole spacing that will be used ☐
(d) 18.1 lb at the minimum pole spacing that will be used ☐

Q7. ASME section V article 7: magnetic particle testing T-752.2

Which types of magnetizing current can be used with the prod technique?

(a) AC or DC ☐
(b) DC or rectified ☐
(c) DC only ☐
(d) They can all be used ☐

Q8. ASME section V article 7: magnetic particle testing T-752.3

What is the maximum prod spacing permitted by ASME V?

(a) It depends on the current being used ☐
(b) There is no maximum specified in ASME codes ☐

(c) 8 in ☐
(d) 6 in ☐

Q9.   ASME section V article 7: magnetic particle testing T-755.1

What is the best description of the limitations of yoke techniques?

(a) They must only be used for detecting surface breaking
    discontinuities ☐
(b) They can also be used for detecting subsurface
    discontinuities ☐
(c) Only AC electromagnet yokes will detect subsurface
    discontinuities ☐
(d) They will detect linear defects in austenitic stainless steels ☐

Q10.   ASME section V article 7: magnetic particle testing: appendix 1

Which MT technique(s) is/are specified in ASME article 7: mandatory appendix 1 to be used to test coated ferritic materials?

(a) AC electromagnet ☐
(b) DC electromagnet ☐
(c) Permanent magnet ☐
(d) AC or DC prods ☐

11.9   ASME V article 10: leak testing

This article of ASME V gives methodologies for *leak testing*. With reference to API 570/ASME B31.3, it provides general requirements to follow when carrying out the leak test. As with the other articles of ASME V, it does *not* tell you when a leak test is or is not required (i.e. in 570 or B31.3). A lot of the content of article 10 is not actually in the API 570 exam syllabus, which makes things easier. The actual amount that *is* in the syllabus is about 5 pages.

T-1010: scope
This article describes various methods of leak testing.

The specific test methods or techniques and glossary of terms of the methods in this article are described in mandatory appendices I to X and non-mandatory appendix A as follows:

Appendix I	Bubble test – direct pressure technique. **This is the only appendix that is in the API syllabus**
Appendix II	Bubble test – vacuum box technique
Appendix III	Halogen diode detector probe test
Appendix IV	Helium spectrometer test – detector technique
Appendix V	Helium spectrometer test – tracer technique
Appendix VI	Pressure change test
Appendix VII	Glossary of terms
Appendix VIII	Thermal conductivity detector probe test
Appendix IX	Helium mass spectrometer test – hood technique
Appendix X	Ultrasonic leak detector test
Appendix A	Supplementary leak testing formula symbols

This all looks very daunting, but for exam purposes you can ignore appendix II onwards. This leaves only the main body of article 10 (4 pages containing general leak testing information) and mandatory appendix I (containing specific details of the bubble test – direct pressure technique) to look at.

T-1040 to T-1052: miscellaneous requirements
Read these short sections in your code, noting the following points, many of which also appear in B31.3:

- Ensure components are dry before leak testing.
- Ensure openings are sealed using materials easily removed after testing.
- Ensure minimum/maximum test temperatures are adhered to.
- Maximum pressure for leak testing is 25 % of design pressure.
- A preliminary test to detect and eliminate gross leaks may be done.
- A leak test is recommended before hydrostatic or hydropneumatic testing.

T-1070: bubble test procedure

See mandatory appendix 1 of this article (i.e. I-1070 to I-1077 of mandatory appendix I for the bubble test – direct pressure technique). Remember that this appendix 1 *is* in the API exam syllabus.

Note how mandatory appendix I contains the following points:

- The test gas is normally air but other inert gases may be used.
- The correct bubble forming solution should be used (e.g. 'Snoop').
- An immersion bath may be used instead of the bubble solution.
- The part temperature should be between 40 °F (4 °C) and 125 °F (52 °C).
- The bubble solution should be applied by brushing, spraying or flowing.

Do not confuse this leak testing with hydrostatic or pneumatic testing. Leak testing is only looking for *leakage* of the component or system and is carried out at pressures well below the design pressure. Hydrostatic or pneumatic testing is carried out at pressures *in excess* of the design pressure and is therefore also a basic strength test.

T-1080: Evaluation of leak test results

T-1081: acceptance standards

Here's some code-speak for you. The code says that:

Unless otherwise specified in the referencing Code Section, the acceptance criteria given for each method or technique of that method shall apply. The supplemental leak testing formulas for calculating leakage rates for the method or technique used are stated in the Mandatory Appendices of this Article.

This means that there should be no bubbles seen when doing a bubble leak test (because I-1081 of mandatory appendix I states that the area under test is acceptable when no continuous bubble formation is observed).

11.10   ASME V article 23: ultrasonic thickness checking

In the ASME V code, this goes by the grand title of *Standard practice for measuring thickness by manual ultrasonic pulse-echo contact method: section SE-797.2*. This makes it sound much more complicated than it actually is. Strangely, it contains some quite detailed technical requirements comprising approximately 7 pages of text and diagrams at a level that would be appropriate to a UT qualification exam. The underlying principles, however, remain fairly straightforward. We will look at these as broadly as we can, with the objective of picking out the major points that may appear as closed-book questions in the API examinations.

The scope of article 23 section SE-797

The technique is for measuring the thickness of any material in which ultrasonic waves will propagate at a constant velocity and from which back reflections can be obtained and resolved. It utilizes the *contact pulse echo method* at a material temperature not to exceed 200 °F (93 °C). Measurements are made from one side of the object, without requiring access to the rear surface.

The idea is that you measure the velocity of sound in the material and the time taken for the ultrasonic pulse to reach the back wall and return (see Fig. 11.7). Halving the result gives the thickness of the material.

Summary of practice

Material thickness (T), when measured by the pulse-echo ultrasonic method, is a product of the velocity of sound in the material and one half the transit time (round trip) through the material. The simple formula is:

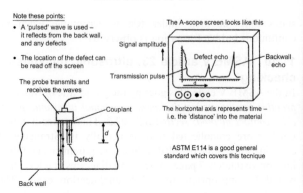

Note these points:

- A 'pulsed' wave is used – it reflects from the back wall, and any defects
- The location of the defect can be read off the screen

The probe transmits and receives the waves

Couplant

Defect

Back wall

The A-scope screen looks like this

Signal amplitude

Transmission pulse

Defect echo

Backwall echo

The horizontal axis represents time – i.e. the 'distance' into the material

ASTM E114 is a good general standard which covers this tecnique

It simply uses the time it takes a pulsed sound wave to pass through a material to give a measure of its thickness

Figure 11.7 UT thickness testing

$$T = \frac{Vt}{2}$$

where T = thickness, V = velocity and t = transit time.

Thickness-checking equipment

Thickness-measurement instruments are divided into three groups:

Flaw detectors with CRT readouts. These display time/amplitude information in an *A-scan* presentation (we saw this method in a previous module). Thickness is measured by reading the distance between the zero-corrected initial pulse and first-returned echo (back reflection), or between multiple-back reflection echoes on a calibrated base line of a CRT. The base line of the CRT should be adjusted to read the desired thickness increments.

Flaw detectors with CRT and direct thickness readout. These are a combination pulse ultrasound flaw detection instrument with a CRT and additional circuitry that provides digital thickness information. The material thickness can be electronically measured and presented on a digital readout. The CRT provides a check on the validity of the electronic measurement by revealing measurement variables, such as internal discontinuities or echo-strength variations that might result in inaccurate readings.

Direct thickness readout meters. Thickness readout instruments are modified versions of the pulse-echo instrument. The elapsed time between the initial pulse and the first echo or between multiple echoes is converted into a meter or digital readout. The instruments are designed for measurement and direct numerical readout of specific ranges of thickness and materials.

Standardization blocks

Article 23 goes into great detail about different types of 'search units'. Much of this is too complicated to warrant too much attention. Note the following important points.

Section 7.2.2.1: calibration (or standardization) blocks
Two 'calibration' blocks should be used: one approximately the *maximum* thickness that the thickness meter will be measuring and the other the *minimum* thickness.

Thicknesses of materials at high temperatures up to about 540 °C (1000 °F) can be measured with specially designed instruments with high temperature compensation. A rule of thumb is as follows:

- A thickness meter reads 1 % too high for every 55 °C (100 °F) above the temperature at which it was calibrated. This correction is an average one for many types of steel. Other corrections would have to be determined empirically for other materials.
- **An example**. If a thickness meter was calibrated on a piece of similar material at 20 °C (68 °F), and if the reading was obtained with a surface temperature of 460 °C (860 °F), the apparent reading should be reduced by 8 %.

SECTION IV: PRESSURE DESIGN

Chapter 12

ASME B31.3: Pressure Design

12.1 B31.3 introduction

This module is about starting to become familiar with the content of ASME B31.3: *Process piping*. This is, basically, a construction code for the manufacture and testing of new pipework. It is subdivided into a large number of chapters and sections, several of which cover subjects relevant to repair, pressure testing and related in-service activities. ASME B31.3 is one of the related codes referenced frequently in API 570 and so forms an important part of the API 570 examination syllabus.

ASME B31.3 is a large document, comprising over 150 pages of closely written text and diagrams. Fortunately, as it is predominantly a standard for new construction, much of this is not included in the API 570 syllabus and less than 30 pages need to be studied in detail. This is supplemented by about a further 20–30 pages that contain more general information about piping systems, a general familiarity with which is useful to help you understand the terminology and general approach of the code.

As with API 570, the ASME B31.3 code cross-references other standards. The main one is ASME B16.5, which covers the design and testing of bolted flanges. Other references are made to the ASME IX (welding) and ASME V (NDE) codes.

Which parts of ASME B31.3 have to be studied for the API 570 exam?

These are fairly well defined and not that difficult to understand, once you have got used to their ideas and terminology. They cover areas such as weld joint efficiency factors, calculation of maximum allowable working pressure

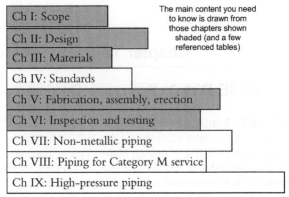

Figure 12.1   ASME B31.3: important sections

(MAWP), pressure testing and repair-related activities such as welding and NDE. This is supplemented by sections covering important mechanical material properties; mainly tensile strength and impact (Charpy) strength. Figure 12.1 shows the main sections of interest.

12.2   B31.3 responsibilities

As a construction code B31.3 has a slightly different viewpoint on the allocation of responsibilities than does API 570. There is little real difference in the substance, however; B31.3 refers to the pipework manufacturer rather than the repair contractor, who is more relevant to the scope and purpose of API 570.

12.3   B31.3 fluid service categories

ASME B31.3 takes a different view to API 570 on the way that process fluids are divided into categories. Whereas API 570 divides systems into risk classes 1, 2, 3 based purely on the consequences of failure, B31.3 takes the approach of user-defined fluid categories based more on the severity of the service. Only one fluid service category (category D, low-risk fluid) is actually prescriptively defined.

The B31.3 fluid categories are (section 300.2):

- Category D (low risk)
- Normal service (the default category)
- Category M service (for toxic fluid)
- High-pressure service

Detailed knowledge of the design difference implicated in pipework desired to the various fluid categories is not a significant part of the API 570 ICP syllabus. A general appreciation of the effect on, for example, the pressure testing option is, however, useful (section B31.3 section 345).

12.4 Pipe wall thickness equations

The API 570 syllabus requires candidates to be able to calculate pipe wall thickness (t_{min}) and MAWP values. These are mainly of use in assessing whether corroded pipework can be safely endorsed for use until the next planned inspection rather than for detailed 'design' calculation, as such.

The equations are found in B31.3 section 304; these cover straight pipes under internal pressure. The main ones (sometimes referred to as the Boardman equations) are, from section 304.1.2:

Wall thickness $(t) = \dfrac{PD}{2(SE + PY)}$ (known as equation 3(a))

and

$t = \dfrac{P(d + 2c)}{2[SEW - P(1 - Y)]}$ (known as equation 3(b))

where

P = internal gauge pressure
D = pipe outside diameter
S = material allowable stress from B31.3 table A-1
E = a longitudinal 'quality factor' from B31.3 table A-1A or A-1B

Y = a coefficient from B31.3 table 304.1.1
W = a weld factor from B31.3 para. 302.3.5(e)
d = pipe inside diameter
c = sum of 'mechanical allowances'

Note the predominance of 'factors' in these equations. While they no doubt have a justified place in the code, many of them (mainly those in equation 3(b)) have limited use in many inspection-related calculations and do not appear regularly in API examination questions. Note how the factor W (only recently added to the code in 2004) only appears in equation 3(b).

Examination questions (open-book) on this topic consist mainly of simple substitution of numbers into these formulae. Variations on the theme include:

- Using a transposed formula to find P (MAWP) when t is given.
- Calculating MAWP at a future time when the current (t) has been reduced by a given 'corrosion allowance'.
- Calculating the safe time to the next inspection on the basis of the 'half-life' principle of API 570.

One interesting point to note is how ASME and API codes take slightly different approaches to wall thickness/MAWP calculations. While B31.3 quotes the Boardman equations 3 (a) and 3(b), API 570/574 prefer the simplified 'Barlow' equation as follows:

$$t = \frac{PD}{2SE} \quad \text{(see API 574 section 11)}$$

In practice, the ASME and API approaches give answers that are fairly close, as long as the design temperature of the pipe is not too high (when the Y-factor of B31.3 table 304.1.1 will have more of an effect).

12.5 B31.3 allowable material stresses

In common with other US codes, B31.3 uses the concept of an *allowable stress* (*S*) for each construction material. The idea of allowable stress is that it is code-defined and chosen to be well below the minimum specified yield strength for a material, thereby introducing a margin of safety into the design. For standard (i.e. non-cryogenic) materials, allowable stress (*S*) decreases as temperature increases, because materials get weaker as they get hotter (something to do with the molecules moving around more and having weaker bonds between them).

Table A-1 of ASME B31.3 comprises a long series of 40 + pages containing all the necessary information on allowable stresses. Finding information from this table is an examinable topic. While not particularly difficult, it can be confusing, because of the length of the table and the many similarly named materials and grades included in it. The objective is normally to find an *S* value, applicable at a given temperature, which is then used in the Boardman or Barlow equations.

12.6 B31.3 impact test requirements

A common theme of API examinations is the question of whether or not materials used (in this context for repairs) require impact testing to see if they have sufficient *toughness* to avoid brittle fracture in service. API and ASME philosophy on this is fairly consistent, but does not necessarily match that used in European and other codes around the world.

Simplistically, the minimum temperature at which a material can be used (i.e. designed to) without impact tests being done is given in B31.3 table A-1. In some cases, instead of giving a single temperature, the table cross-references section 323 of the code. Materials are divided into four groups A, B, C and D, and their minimum design temperature (without needing impact tests) is read off the relevant graph (see figure 323.2.2A as an example). Look at this graph and you will see how the minimum temperature varies with nominal thickness of the material (thick material is more brittle) and whether or not it is normalized (which refines the grain structure, reducing brittleness).

The situation is complicated slightly by the code principle that a pipe material which is under particularly low stress is allowed further reductions in its design temperature without impact tests. Have a look at figure 323.2.2.B and you will see how this works.

12.7   ASME B31.3 familiarization questions

Now try these ASME B31.3 familiarization questions.

Q1.   B31.3 section 300: fluid service groupings

How many different fluid groupings are mentioned in chapter 1 of B31.3?

(a) 2 □
(b) 3 □
(c) 4 □
(d) 8 □

Q2.   B31.3 section 300: fluid service groupings

Which of these is the lowest risk fluid class?

(a) Category M □
(b) Category A □
(c) Category D □
(d) 'Normal' class □

Q3.   B31.3 section 300: fluid service groupings

A dangerous toxic fluid such as cyanide gas would be in which fluid category?

(a) Category A □
(b) Category D □
(c) Category M □
(d) Category X □

Q4.   B31.3 section 300.1.3: exclusions

Which of these would be *excluded* from the coverage of B31.3?

(a) A small pressure vessel □
(b) An NPS 2 pipe with a MAWP of 1.3 barg □

(c) A plastic pipe designed for 20 psi compressed air ☐
(d) A copper pipe designed for 20 psi cold water ☐

Q5. B31.3 section 300.1.3: exclusions

Which of these would be *within* the scope of B31.3 (i.e. not excluded)?

(a) A pipe designed to contain water at 90 °C, 90 kPa ☐
(b) A pipe designed to contain non-hazardous chemical at 200 °C, 0.5 barg ☐
(c) Steam pipes between a high-pressure boiler and steam turbine ☐
(d) A shell-and-tube heat exchanger designed for 200 kPa steam ☐

Q6. B31.3 section 302.3.1

What data are included in B31.3 table A-1?

(a) Weld joint factors ☐
(b) Casting quality factors ☐
(c) Allowable stresses in tension for metals ☐
(d) Values for the mysterious coefficient Y ☐

Q7. B31.3 section 302.3.2: strength symbols

The generic symbol used for stress and strength in API/ASME codes is:

(a) S ☐
(b) T ☐
(c) σ (Greek sigma) ☐
(d) Y ☐

Q8. B31.3 section 302.3.2

Where would you find a table of casting quality factors in B31.3?

(a) Table A-1 ☐
(b) Table A-1A ☐
(c) Table A-1B ☐
(d) Table 302.3.4 ☐

Q9. B31.3 section 302.3.2: strength symbols

What does the symbol S_T mean in B31.3?

(a) Yield strength at the design temperature ☐
(b) Ultimate tensile strength (UTS) at the design temperature ☐
(c) Yield strength at room temperature ☐
(d) Ultimate tensile strength (UTS) at room temperature ☐

Q10. B31.3 section 302.3.2: strength symbols

What does the symbol S_y mean in B31.3?

(a) Yield strength at the design temperature   ☐
(b) Ultimate tensile strength (UTS) at the design
    temperature   ☐
(c) Yield strength at room temperature   ☐
(d) Ultimate tensile strength (UTS) at room temperature   ☐

Q11. B31.3 section 302.3.3: casting quality factor

What is the symbol used in B31.3 for casting quality factor?

(a) C_f ☐
(b) E ☐
(c) E_c ☐
(d) E_j ☐

Q12. B31.3 section 302.3.3: weld quality factor

The primary purpose of allocating a weld quality factor to a longitudinal pipe weld is to take into account which one of these?

(a) The increased stress on a weld   ☐
(b) Limitations on the NDE performed on the weld   ☐
(c) The increased strength of the weld over the parent
    material   ☐
(d) Lack of documentation about the welding procedure   ☐

Q13. B31.3 section 304.1.1: coefficient Y

The coefficient Y is used in the formula for calculation of the required wall thickness of a straight length of pipe under internal pressure. Why is Y actually there?

(a) To take into account the reduction in strength of welds   ☐
(b) As an allowance for shock loads   ☐
(c) Because pipes vary in their t/D ratio   ☐
(d) To cater for temperatures below ambient   ☐

Q14. B31.3 section 304.1.1: coefficient Y

For a pipe of wall thickness t, the cut-off point for the ratio of outside diameter D to inside diameter d at which the source data for finding Y changes is:

(a) $d = 10t$ ☐
(b) $d = 6t$ ☐
(c) $D = 10t$ ☐
(d) $D = 6t$ ☐

Q15.   B31.3: allowable stresses: table A-1

Table A-1 covers allowable stresses in pipework materials. The table displays its units of stress and temperature in:

(a) ksi and °F only ☐

(b) kPa and °F only ☐

(c) ksi/MPa and °F ☐

(d) ksi/MPa and °F/°C ☐

Chapter 13

ASME B16.5 Flange Design

13.1 Introduction to ASME B16.5

This chapter is about starting to become familiar with the content of ASME B16.5: *Pipe flanges and flanged fittings*. ASME B16.5 is a construction code for the design of new flanges and related fittings such as reducers, tees and similar. It is one of the related codes referenced in API 570 and API 574 and so forms an important (but not particularly large) part of the API 570 examination syllabus.

Like ASME B31.3, ASME B16.5 is a large document, comprising over 200 pages. Fortunately it is much easier to understand than B31.3 as it contains very few formulae and symbols. Its main content is 10 pages or so of specific design requirements at the front of the code (sections 1 to 8) followed by multiple data tables containing materials and temperature/pressure ratings for various types of flanges. This is supplemented by appendices that contain more specific information about piping systems, none of which is directly required by the API 570 syllabus.

13.1.1 Which parts of ASME B16.5 have to be studied for the API 570 exam?

The study required of B16.5 is much easier than that for B31.3. API state that the main tasks that API 570 examination candidates must be able to do for pipe flanges are as follows:

- Find the minimum wall thickness and working pressure requirements.
- Find the working pressure and minimum/maximum system hydrostatic test pressure.
- Find the minimum dimensions of a given flange.
- Find the maximum working pressure of a flange when

given the design temperature, flange material and flange class.

- Find the maximum temperature of a flange when given the design pressure, flange material and flange class.
- Find the most cost effective flange when given the design pressure, design temperature, and flange material.

This looks a bit complicated but here's the good news: *in practice, all of this amounts to little more than reading from tables in the code.* There are lots of tables, but only because there are many different flange sizes, pressure classes and materials. In essence, the tables are all much the same – just repeated many times with different data in them. Compared to ASME B31.3 and API 570, there is very little of the code that you actually have to learn; it is more a case of just knowing which tables to look at and how to interpret the data without making mistakes.

13.2 Familiarization questions: ASME B16.5 flange design (see Fig. 13.1)

Try these familiarization questions.

Q1. What type of flange is flange A?
(a) An RTJ flange ☐
(b) A slip-on flange ☐
(c) A lapped-end flange ☐
(d) A socket-weld flange ☐

Q2. What type of flange is flange B?
(a) An RTJ flange ☐
(b) A slip-on flange ☐
(c) A lapped-end flange ☐
(d) A socket-weld flange ☐

Q3. What type of flange is flange C?
(a) A socket weld flange ☐
(b) A lapped-end flange ☐
(c) A threaded flange ☐
(d) A weld-neck flange ☐

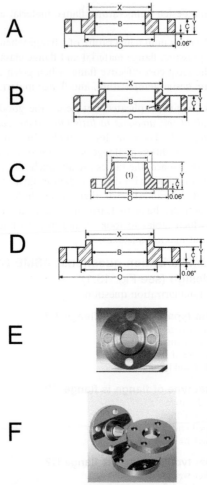

Figure 13.1   ASME B16.5 flange types

Q4. What is the maximum size of flange covered by ASME B16.5?

(a) NPS 12 ☐
(b) NPS 20 ☐
(c) NPS 24 ☐
(d) NPS 36 ☐

Q5. What does a #150 flange class designation mean?

(a) The maximum flange working pressure is 150 psi ☐
(b) The maximum flange test pressure is 150 psi ☐
(c) The maximum flange working pressure is 150 psi at 100 °F ☐
(d) None of the above ☐

Q6. Which parameter in Fig. 13.1 for flange D represents the flange *wall thickness*?

(a) c ☐
(b) $(R - B)/2$ ☐
(c) $(x - B)/2$ ☐
(d) $(R - x)/2$ ☐

Q7. Which of these flanges made of the same material would you expect to have the highest safe working pressure at 200 °F?

(a) #150 NPS 4 ☐
(b) #300 NPS 4 ☐
(c) #1500 NPS 4 ☐
(d) #2500 NPS 4 ☐

Q8. What is the format of the pressure units used in ASME B16.5?

(a) psig (kPa) ☐
(b) kPa (psig) ☐
(c) psig and barg ☐
(d) psig only ☐

Q9. What type of flange facing is flange E in Fig. 13.1?

(a) An RTJ flange ☐
(b) An RF flange ☐
(c) A flat-face flange ☐
(d) A tongue and groove flange ☐

Q10.   Which three flange types are shown in Fig. 13.1 for flange F?

(a) Threaded, blind and socket ☐

(b) Socket, weld neck and lapped ☐

(c) Blind, lapped and weld neck ☐

(d) Weld neck, socket and blind ☐

SECTION V: EXAMPLE QUESTIONS

Chapter 14

Example Open-Book Questions

Try these questions, using your codes to find the necessary formulae or references. The answers are given at the end of this book.

Question 1
What is the minimum design temperature for carbon steel pipe 20 mm thick with a design pressure of 100 psig containing a fluid which is non-flammable, non-toxic and not damaging to human tissue?

(a) 15 °C ☐
(b) −8 °C ☐
(c) −29 °C ☐
(d) −25 °C ☐

Question 2
What is the maximum allowable working pressure for a 20 in standard wall pipe made in seamless material of allowable stress 20 ksi operating at ambient temperature?

(a) 650 psi ☐
(b) 700 psi ☐
(c) 750 psi ☐
(d) 800 psi ☐

Question 3
What is the maximum non-shock pressure rating at 250 °C for a class 1500 flange made of A350 LF 2 material?

(a) 3705 psig ☐
(b) 3750 psig ☐
(c) 3041 psig ☐
(d) 2490 psig ☐

Question 4
What is the upper temperature limit for carbon steel to operate without degradation?

(a) 600 °F ☐
(b) 700 °F ☐
(c) 800 °F ☐
(d) 850 °F ☐

Question 5
What is the minimum thickness required for a 10 in ERW pipe in API 5L Gr.B operating at 260 °C with a corrosion allowance of 2 mm operating at 348 psig? Round up your answer to the nearest mm.

(a) 3 mm ☐
(b) 4 mm ☐
(c) 5 mm ☐
(d) 6 mm ☐

Question 6
You have manufactured a pipe from plate made to a material which is not included in tables 1A and A-1 of B31.3. The mill certificate for the steel gives the yield strength of 2070 kPa and an ultimate tensile strength of 4300 kPa. The pipe operates at ambient temperature. What is the value of the design stress?

(a) 1380 kPa ☐
(b) 1430 kPa ☐
(c) 1075 kPa ☐
(d) 1863 kPa ☐

Question 7
A cast material has been provided in A216 WCB material for which all the surfaces have been machined to an undefined surface finish, MT tested, and no defects found. What casting factor is used in calculating the required thickness?

(a) 0.65 ☐
(b) 0.85 ☐
(c) 0.90 ☐
(d) 1.00 ☐

Question 8. B31.3
An 8 in class 600 flange in A182 F304L material operates at 400 °C at 670 psi. If the gasket diameter is 278 mm, what thickness of blank is required?

(a) 20 mm ☐
(b) 25 mm ☐
(c) 27 mm ☐
(d) 30 mm ☐

Question 9

A class 300 flange made of A182 F11 material has a pressure limit of 4630 kPa. What is the maximum temperature at which this flange can operate?

(a) 200 °C ☐
(b) 250 °C ☐
(c) 260 °C ☐
(d) 500 °C ☐

Question 10

A class 150 flange in A105 material has the design conditions of 170 psi at 500 °F. If the gasket diameter is 254 mm, what thickness of blank is required?

(a) 10 mm ☐
(b) 15 mm ☐
(c) 20 mm ☐
(d) 25 mm ☐

Question 11

What is the minimum design temperature for a pipe made in A516 Gr55 containing liquid petroleum gas at 150 psig that is 30 mm thick?

(a) 0 °C ☐
(b) −10 °C ☐
(c) −15 °C ☐
(d) −20 °C ☐

Question 12

If a material has a design stress of 130 psig and the pressure inside the pipe produces a stress of 62 psig, what reduction can be made in the minimum design temperature without impact testing the material?

(a) 10 °C ☐
(b) 26 °C ☐
(c) 36 °C ☐
(d) 45 °C ☐

Question 13

If the material has a design stress of 160 psig and the internal pressure that produces a stress of 64 psig, what reduction can be made in the minimum design temperature without impact testing the material?

(a) 20 °C ☐
(b) 40 °C ☐
(c) 60 °C ☐
(d) 80 °C ☐

Question 14

What is the *minimum* value of charpy impact required for steel made of fully deoxidized API 5L Gr.B pipe?

(a) 18 joules ☐
(b) 16 joules ☐
(c) 10 joules ☐
(d) 7 joules ☐

Question 15

A pipe made in API 5L Gr.B operates at 260 °C and is 22.86 metres long. How much will the pipe expand with an ambient temperature of 70 °F?

(a) 69 mm ☐
(b) 59 mm ☐
(c) 79 mm ☐
(d) 54 mm ☐

Question 16

A pipe made in A312 304 material operates at 510 °C and is 45.72 metres long. How much will the pipe expand with an ambient temperature of 70 °F?

(a) 310 mm ☐
(b) 411 mm ☐
(c) 523 mm ☐
(d) 157 mm ☐

Question 17
An 18 in standard wall thickness pipe has been made by furnace butt welding from a piece of A515 Gr 60 plate. It operates at 600 °F at 300 psi. If the corrosion rate has been measured at 0.254 mm per year, when should the next planned inspection take place?

(a) 2.7 years ☐
(b) 3.9 years ☐
(c) 4.7 years ☐
(d) 6 years ☐

Question 18
If the actual thickness of a pipe is measured at 10 mm and the design thickness is 6 mm and there is a corrosion rate of 0.25 mm per year, when should the pipe be replaced?

(a) 10 years ☐
(b) 16 years ☐
(c) 24 years ☐
(d) 30 years ☐

Question 19
A thin-walled stainless steel pipe of 20 in diameter by 3 mm has a design pressure of 254 psig and operates at ambient temperature. What is the hydraulic test pressure?

(a) 381 psig ☐
(b) 300 psig ☐
(c) 254 psig ☐
(d) 428 psig ☐

Question 20
What is the corrosion rate for a pipe 10 in Sch 140 thickness that has a measured thickness of 15.4 mm after 10 years of service?

(a) 0.5 mm/year ☐
(b) 1 mm/year ☐
(c) 1.5 mm/year ☐
(d) 2 mm/year ☐

Question 21
In the year 2000, thickness readings were taken from a 6 in Sch 40 pipe which gave a minimum reading of 7 mm. The measurements were repeated in 2004 and this time the minimum reading was 5 mm. What is the corrosion rate?

(a) 0.25 mm/year ☐
(b) 0.50 mm/year ☐
(c) 0.75 mm/year ☐
(d) 1.25 mm/year ☐

Question 22
What is the 'Barlow' formula used for?

(a) To determine the corrosion life of the pipe ☐
(b) To determine the retirement thickness of the pipe ☐
(c) To determine the expansion of the pipe ☐
(d) To find the temperature reduction for non-impact tested materials ☐

Question 23
What is the minimum wall thickness for a valve attached to a pipe that has a design thickness of 10 mm?

(a) 7.5 mm ☐
(b) 10 mm ☐
(c) 15 mm ☐
(d) 20 mm ☐

Question 24
A flanged fitting has a measured thickness of 30 mm. What is the *minimum* thickness of pipe that should be attached to this?

(a) 30 mm ☐
(b) 20 mm ☐
(c) 15 mm ☐
(d) 45 mm ☐

Question 25
How often should thickness measurements be made on a pipe system containing ethylene gas?

(a) Every 2 years ☐
(b) Every 3 years ☐
(c) Every 5 years ☐
(d) Every 10 years ☐

Question 26
How often should thickness measurements be made on a pipe system containing hydrogen?

(a) Every 2 years ☐
(b) Every 3 years ☐
(c) Every 5 years ☐
(d) Every 10 years ☐

Question 27
How often should external visual inspection of a pipe system containing hydrofluoric acid be carried out?

(a) Every 2 years ☐
(b) Every 3 years ☐
(c) Every 5 years ☐
(d) Every 10 years ☐

Question 28
A pipeline carrying caustic soda runs off-site and is lagged. The lagging is in poor condition. What percentage of the pipe should be examined?

(a) 10 % ☐
(b) 25 % ☐
(c) 33 % ☐
(d) 50 % ☐

Question 29
A lagged carbon pipe system containing propylene gas operates at 85 °C. What percentage of the pipe should be examined?

(a) 10 % ☐
(b) 25 % ☐
(c) 33 % ☐
(d) 50 % ☐

Question 30
A $\frac{1}{2}$ in pipe is connected to a line containing natural gas as a secondary process system. What is the interval for visual inspection on the pipe?

(a) Every 5 years ☐
(b) Every 2 years ☐
(c) Every 10 years ☐
(d) Inspection is optional ☐

Chapter 15

Answers

15.1 Familiarization answers

Subject question and Chapter	Question number	Answer
API 570 (Chapter 3)	1	a
	2	b
	3	c
	4	d
	5	c
	6	b
	7	d
	8	b
	9	b
	10	c
	11	b
	12	a
	13	d
	14	d
	15	d
API 574 (Chapter 4)	1	c
	2	a
	3	b
	4	c
	5	c
API 574 (sections 6 and 10) (Chapter 4)	1	a
	2	c
	3	d
	4	b
	5	b
	6	d
	7	b
	8	c
	9	d
	10	c
API 578 (Chapter 5)	1	b
	2	c
	3	c

Answers

Subject question and Chapter	Question number	Answer
	4	a
	5	b
	6	b
	7	c
	8	b
	9	d
	10	a
API 571 (set 1) (Chapter 6)	1	b
	2	c
	3	c
	4	a
	5	c
	6	b
	7	c
	8	b
	9	b
	10	b
API 571 (set 2) (Chapter 6)	1	b
	2	b
	3	a
	4	b
	5	d
	6	d
	7	d
	8	d
	9	b
	10	b
API 571 (set 3) (Chapter 6)	1	d
	2	a
	3	c
	4	a
	5	b
	6	c
	7	b
	8	b
	9	d
	10	d
API 577: welding process (Chapter 7)	1	a
	2	b
	3	b

Subject question and Chapter	Question number	Answer
	4	a
	5	d
	6	d
	7	c
	8	c
	9	b
	10	b
API 577: welding consumables (Chapter 7)	1	a
	2	b
	3	b
	4	a
	5	a
	6	d
	7	c
	8	c
	9	b
	10	d
API 570: general welding rules (Chapter 8)	1	a
	2	b
	3	c
	4	d
	5	a
	6	b
	7	d
	8	d
	9	b
	10	d
ASME B31.3 (Chapter 8)	1	a
	2	b
	3	c
	4	b
	5	a
ASME IX articles I and II (Chapter 9)	1	c
	2	c
	3	d
	4	a
	5	c
	6	a
	7	b
	8	b

Subject question and Chapter	Question number	Answer
	9	d
	10	c
ASME IX articles III and IV (Chapter 9)	1	d
	2	d
	3	d
	4	b
	5	a
	6	a
	7	a
	8	a
	9	c
	10	d
API 577 (Chapter 10)	1	c
	2	c
	3	b
	4	d
	5	a
	6	b
	7	c
	8	d
	9	d
	10	d
API 570 and ASME B31.3(1) (Chapter 10)	1	a
	2	d
	3	d
	4	a
	5	b
	6	c
	7	a
	8	a
	9	c
	10	b
ASME B31.3 (2) (Chapter 10)	1	a
	2	d
	3	b
	4	d
	5	a
	6	b
	7	c
	8	a

Subject question and Chapter	Question number	Answer
	9	d
	10	c
ASME V articles 1, 2 and 6 (Chapter 11)	1	d
	2	d
	3	d
	4	d
	5	c
ASME V articles 7 and 9 (Chapter 11)	1	b
	2	a
	3	d
	4	d
	5	d
	6	a
	7	b
	8	c
	9	a
	10	a
ASME B31.3: pressure design (Chapter 12)	1	b
	2	c
	3	c
	4	a
	5	b
	6	c
	7	a
	8	b
	9	d
	10	c
	11	c
	12	b
	13	c
	14	d
	15	d
ASME B16.5: flange design (Chapter 13)	1	b
	2	c
	3	d
	4	c
	5	d
	6	a
	7	d
	8	c

Subject question and Chapter	Question number	Answer
	9	b
	10	c

15.2 Example open-book answers

Question 1. B31.3: minimum design temperatures

B31.3 graph 323.2.1, as this is a Category D fluid. Note 1 of the table says that any carbon steel can be used down to –29 °C (ANS) (as long as it's a Category D fluid).

Question 2. B31.3: Barlow equation

API 570 7.2 and using the simple Barlow equation $P = 2SE\,t/D$ for the MAWP calculations,

$E = 1$ for seamless, wall thickness t for a 20 in std pipe (API 574) = 0.375 in, OD (D) of pipe = 20 in, $P = 2 \times 20\,000 \times 0.375/20$ = **750 psi (ANS)**.

Question 3. Flange test pressures

B16.5 table 1A and table 2 show that A350 LF2 material is table 1-1. At 250 °C, a Cl 1500 flange has a maximum working pressure of 209.7 bar = **3041.2 psi (ANS)**.

Question 4. Degradation mechanisms

API 574 section 10.2.1.5.2: suffers graphitization above **800 °F** (ANS).

Question 5. B31.3: wall thickness equations

B31.3 section 304.1.2 equation 3(a). Using $t = PD/[2(SE + PY)]$ in equation 3(a), note that for a 10 in NB pipe, the OD is 10.75 in, not 10 in; $S = 18.9$ ksi for API 5L at 260 °C (500 °F), $Y = 0.4$ for ferritic steel (table 304.1.1), $E = 0.85$ for ERW pipe from table 302.3.4, $t = (348 \times 10.75)/\{2 \times [(18\,900 \times 0.85) + (348 \times 0.4)]\} = 0.115$ in = 2.93 mm + 2 mm rounded up = **5 mm (ANS)**.

Question 6. B31.3: design stress

From the 'other materials' section of B31.3, 302.3.2 (d), the design stress shall not exceed:

- one-third of the minimum UTS at room temperature (S_T) or at temperature;

- two-thirds of the minimum yield at room temperature (S_Y) or at temperature,

so in this case it's the lower of 4300 kPa/3 = 1433 kPa or $\frac{2}{3} \times 2070$ kPa = **1380 kPa (ANS)**.

Question 7. B31.3: casting factors
Read from the table in B31.3 section 302.3.3c. The answer is **0.85 (ANS)** as no surface finish is given in the question.

Question 8. B31.3: blank thickness
Using the blank formula in B31.3 section 304.5.3, $t = d$ SQRT $[3P/(16SE)] + c$, $D = 278$ mm = 10.94 in, $P = 670$ psi, $S =$ A182 F304L material at 400 °C (752 °F) = 13.2 ksi, $t = 10.94 \times$ SQRT $[(3 \times 670)/(16 \times 13\ 200)] = 1.06$ in = **27 mm (ANS)**.

Question 9. Flange test pressures
B16.5 table 1A and table 2-1.9. For a class 300 flange the table shows a maximum temperature of **250 °C (ANS)**.

Question 10. B31.3: blank thickness
Using the blank formula in B31.3 section 304.5.3, $t = d$ SQRT $[3P/(16SE)] + c$, $D = 10$ in, $P = 170$ psi, $S =$ A105 material at 500 °F = 19.4 ksi, $t = 10 \times$ SQRT $[(3 \times 170)/(16 \times 19\ 400)] = 0.4$ in = 10.29 mm; to go for the next size up **15 mm (ANS)**.

Question 11. B31.3: minimum design temperatures
B31.3 graph 323.3.2A. Curve C at 30 mm thick shows **−15 °C (ANS)**.

Question 12. B31.3: minimum design temperature without impact testing
B31.3 graph 323.3.2B. Stress ratio = 62/130 = 0.467 gives a temperature reduction of 60 °F **(36 °C) (ANS)**.

Question 13. B31.3: allowable reductions in design temperature
B31.3 graph 323.3.2B. Stress ratio = 64/160 = 0.4 gives a temperature reduction of **60 °C (ANS)**.

Question 14.   B31.3: minimum Charpy values

Method: B31.3 section 323.3.5. Checking strength from B31.3 table A-1 gives 20 ksi. Looking in table 323.3.5 for 65 ksi and less gives a minimum Charpy value of **16 J (ANS)**.

Question 15.   B31.3: thermal expansion

From B31.3 appendix C table, temperature rise from 70 to 500 °F gives an expansion of carbon steel (API 5L is carbon steel) of 3.62 in/100 ft. Pipe is 900.68 in = 75 feet so expansion = $3.62 \times 0.75 = 2.71$ in = **69 mm (ANS)**.

Question 16.   B31.3: thermal expansion

From B31.3 appendix C table, temperature rise from 70 to 950 °F gives an expansion of stainless steel (API 5L is carbon steel) of 10.8 in/100 ft. Pipe is 1801.36 in = 150 feet so expansion = $10.8 \times 1.5 = 16.2$ in = **411 mm (ANS)**.

Question 17.   API 570: corrosion rates

Method: API 570 section 7.2. 18 in std has wall thickness of 0.375 in = 9.525 mm. A515 Gr 50 has S = 15.8 ksi at 600 °F. E = 0.6 for furnace butt welded pipe Y = 0.4 for ferritic pipe. Using t = PD/[2($SE + PY$)], t = 300×18 /2 [(15 800 × 0.6) + (300 × 0.4)] = 5400/19 200 = 0.28 in = 7.14 mm. Hence it has 9.525 − 7.14 mm = 2.385 mm to corrode at 0.254 mm/yr = 9.4 years. It should be inspected after half of this = **4.7 years (ANS)**.

Question 18.   API 570: corrosion rates

API 570 section 7.1.1. Hence it has 10 − 6 mm = 4 mm to corrode at 0.25 mm/yr = 16 years. It should be replaced after this **16 years (ANS)**.

Question 19.   Flange and pipe test pressures

Method: B31.3 section 345.2.1. Pipe size details are not required. It is simply Pt = 1.5 $P \times S_{test}/S_{temp}$ = 1.5×254 = **381 psi (ANS)**.

Question 20.   API 570: corrosion rates

Method: API 570 section 7.1.1. Check 10 in Sch 140 has a wall thickness of 1 in = 25.4 mm. So it has corroded 10 mm in 10 years = **1 mm/year (ANS)**.

Question 21.   API 570: corrosion rates

Method: API 570 section 7.1.1. So it has corroded 2 mm in 4 years = **0.5 mm/year (ANS)**.

Question 22.   API 574: Barlow formula
Method: API 574 section 11.1. Answer is **(b) (ANS)**.

Question 23.   API 574: valve wall thickness
API 574 section 11.2 specifies that a valve must have **1.5 × wall thickness** of a pipe connected to it. Hence thickness = 1.5×10 mm = **15 mm (ANS)**.

Question 24.   API 574: flanged connection thickness
API 574 section 11.2 specifies that a flanged fitting (like a valve) must have **1.5 × wall thickness** of a pipe connected to it. So if the flanged fitting is 30 mm the pipe must be **20 mm (ANS)**.

Question 25.   API 570: pipe classes
API 570 table 6.1, class 1 pipe. Thickness measurements are required **every 5 years (ANS)**.

Question 26.   API 570: pipe classes
API 570 table 6.1, class 2 for hydrogen. Thickness measurements are required **every 10 years (ANS)**.

Question 27.   API 570: pipe classes
API 570 table 6.1, class 2 for strong acids and caustics. External visual is required **every 5 years (ANS)**.

Question 28.   API 570: pipe classes CUI
Method: API 570 table 6.2 and table 6.2, class 3. Required examination is **25 % (ANS)**.

Question 29.   API 570: pipe classes CUI
API 570 table 6.2, class 1 but no damaged lagging. Required examination is **50 % (ANS)**.

Question 30.   API 570: pipe classes inspection period
API 570 section 6.6.1, class 2 small-bore secondary pipework. Code says **inspection is optional (ANS)**.

Appendix

Publications Effectivity Sheet

For API 570 Exam Administration: 4th June 2008

Listed below are the effective editions of the publications required for the **June 2008** API 570, Piping Inspector Certification Examination.

- **API Standard 570,** *Piping Inspection Code: Inspection, Repair, Alteration, and Rerating of In-Service Piping Systems,* 2nd Edition, October 1998, Addenda 1, 2, 3 **and Addenda 4 (June 2006).**

IHS Product Code **API CERT 570**

- **API Recommended Practice 571**, *Damage mechanisms affecting fixed equipment in the refining industry, 1st Edition, December 2003.* IHS Product Code **API CERT 570_571** (includes only the portions listed below)

ATTENTION: Only the following mechanisms listed in RP 571 to be included:

Par. 4.2.7 – Brittle Fracture
    4.2.9 – Thermal Fatigue
    4.2.14 – Erosion/Erosion Corrosion
    4.2.16 – Mechanical Fatigue
    4.2.17 – Vibration-Induced Fatigue
    4.3.2 – Atmospheric Corrosion
    4.3.3 – Corrosion Under Insulation (CUI)
    4.3.5 – Boiler Water Condensate Corrosion
    4.3.7 – Flue Gas Dew Point Corrosion
    4.3.8 – Microbiologically Induced Corrosion (MIC)
    4.3.9 – Soil Corrosion
    4.4.2 – Sulfidation
    4.5.1 – Chloride Stress Corrosion Cracking (Cl⁻SCC)

4.5.3 – Caustic Stress corrosion Cracking (Caustic Embrittlement)

5.1.3.1 – High Temperature Hydrogen Attack (HTTA)

- **API Recommended Practice 574**, *Inspection Practices for Piping System Components, Second Edition, June 1998.* IHS Product Code **API CERT 574**
- **API Recommended Practice 577** - *Welding Inspection and Metallurgy, 1st edition, October 2004.* IHS Product Code **API CERT 577**
- **API Recommended Practice 578**, *Material Verification Program for New and Existing Alloy Piping Systems, 1st Edition, May 1999.* IHS Product Code **API CERT 578**
- **American Society of Mechanical Engineers (ASME)**, *Boiler and Pressure Vessel Code,* 2004 edition with 2005 addenda **and 2006 addenda**

ASME Section V, *Nondestructive Examination, Articles I, 2, 6, 7, 9, 10, and 23 (Section SE-797 only).*

Section IX, *Welding and Brazing Qualifications, Welding only*

- **American Society of Mechanical Engineers (ASME)**
  i. BI6.5, *Pipe Flanges and Flanged Fittings,* 2003 Edition
  ii. B31.3, *Process Piping,* 2004 Edition
  IHS Product Code for the ASME package **API CERT 570 ASME**. Package includes only the above excerpts necessary for the exam.

API and ASME publications may be ordered through IHS Documents at 303-397-7956 or 800-854-7179. Product codes are listed above. Orders may also be faxed to 303-397-2740. More information is available at http://www.ihs.com. API members are eligible for a 30% discount on all API documents; exam candidates are eligible for a 20% discount on all API documents. When calling to order, please identify yourself as an exam candidate and/or API member. **Prices**

quoted will reflect the applicable discounts. No discounts will be made for ASME documents.

Note: API and ASME publications are copyrighted material. Photocopies of API and ASME publications are not permitted. CD-ROM versions of the API documents are issued quarterly by Information Handling Services and are allowed. Be sure to check your CD-ROM against the editions noted on this sheet.

Index

Printed and bound by CPI Group (UK) Ltd, Croydon, CR0 4YY

15/10/2024

01774590-0001